A Detailed Chronological Record of Project 523 and the Discovery and Development of *Qinghaosu* (Artemisinin)

Editorial Board members of this book. From left: Shi Linrong, Song Shuyuan, Zhang Jianfang, Zhou Keding, Fu Liangshu, Wang Huansheng.

A Detailed Chronological Record of Project 523 and the Discovery and Development of *Qinghaosu* (Artemisinin)

Zhang Jianfang

Strategic Book Publishing and Rights Co.

Strategic Book Publishing and Rights Co.
12620 FM 1960, Suite A4-507
Houston TX 77065
www.sbpra.com

ISBN: 978-1-62212-164-9

EDITOR ZHANG JIANFANG

BEIJING

Principle contributors:

Zhang Jianfang: Former Deputy Director, National Project 523 Head Office

Zhou Keding: Former Key Member, National Project 523 Head Office; Former Member and Secretary *Qinghaosu* Directional Committee

Zhou Yiqing: Former Director, Beijing District Project 523 Office

Shi Linrong: Former Assistant, National Project 523 Head Office

***

Other contributors:

Fu Liangshu: Former Director, Kunming District Project 523 Office

Wang Huansheng: Former Director, Shanghai District Project 523 Office

Song Shuyuan: Researcher, Senior Professor Pharmacology & Toxicology, Military Academy of Medical Science; Scientific and Technology Consultant for this book

***

Zhang Jianfang

Translation from Chinese with editorial revision: Muoi Arnold and Keith Arnold

Note: The use of qinghaosu and/or artemisinin is not arbitrary but related to context.

Cover photos: Plant: *Qinghao (Artemisia annua L)*
Purified extract: *Qinghaosu* (Artemisinin)

Photos in text on pages 100 to 105.

For all those involved in the project

***

Initial planners and the People's Liberation Army;
Field-workers scouring the countryside for herbal remedies;
Scientists in laboratories for analysis of herbal compounds and
synthesis of new compounds;
Biologists with animal models for efficacy and safety of
potential therapeutic agents;
Clinicians for the care of patients receiving experimental
treatments;
Administrators for overseeing the complex organization and
data processing

CONTENTS

ACKNOWLEDGMENTS

The foregoing detailed account is a collation of reports written by former key personnel of the national Project 523 leading group office and local 523 offices. The editor thoroughly reviewed all the reports for content, reliability, and accuracy. His concern has been for historical verisimilitude, which he ensured by examining the original documents and data whenever possible and by interviewing members of the many research units involved in the project. Sites visited included Beijing, Shandong, Shanghai, and Yunnan. Meetings were held with participants in "alumni" gatherings to check document validity and verify authenticity of events described in the submitted reports. The initial draft was reviewed, discussed, and modified many times as the number of interviewees progressively enlarged, and more input was obtained from various leading institutions, departments, units, and professional groups. This corrected draft was then sent for further review and opinion and for any changes to be made, to leaders of the previous Project 523 leading group, leaders and experts in districts and units that participated in Project 523 *qinghaosu* (artemisinin) research and development. In addition, a meeting was organized in Beijing on November 26, 2005 for Beijing residents who were former personnel of the Project 523 leading group and those who had worked with the project, again to obtain input on the draft. It took more than two years to complete the final manuscript as presented here.

Although thirty years have passed, the review and summarizing of the chronology of events of Project 523 and the work on *qinghaosu* still bring back the pride and positive feelings of those involved in the project. To see the enormous benefits that have resulted from the discovery of *qinghaosu*, due to the dedication and hard work of so many is a credit to the collaborative spirit of socialism.

In recent years some newspapers and magazines have stated that *qinghaosu* was the research product of a certain unit or a certain individual, and therefore this unit or individual owns the patent rights of the new drug *qinghaosu*. We all feel that this statement is clearly at variance with the facts and totally inconsistent with the historical record. *Qinghaosu* and its derivatives and combinations are the collective achievement of a nationwide collaborative effort, and beyond dispute.

The title of this monograph was selected by Qian Xinzhong, former minister of health and ex-leading member of the national Project 523 leading group. The introduction was written by Chen Haifeng, former director of the Department of Scientific Education, in the Ministry of Health, member of previous national Project 523 leading group, and the first chairman of the *Qinghaosu* Directional Committee. Many opinions and corrections were provided by former project leaders and specialists, including Cong Zhong, Yang Shuyu, Shen Jiaxiang, Ji Ruyun, Chen Ningqing, Song Shuyuan, Zhou Hai, Luo Zeyuan, Wang Heng, Wu Huizhang, Wan Yaode, Li Ying, Wu Yulin, Zhang Shugai, Li Zelin, Zeng Meiyi, Zhang Kui, Su Fachang, Zhan Eryi, Wang Cunzhi, Li Guoqiao, Wei Jianyun, Guo Anrong, and Meng Heyan. I extend my sincere thanks to all of them.

***

Zhang Jianfang, Editor
Beijing, January 2006

INTRODUCTION

After reading this account of Project 523 which details the discovery and development of *qinghaosu*, edited by Mr. Zhang Jianfang of the National Project 523 Office, I felt that I was reliving those strenuous times of thirty some years ago. Following instructions from Chairman Mao and with approval of the State Council, State Military Commission, and Prime Minister Zhou Enlai to be combat ready and to help our ally, a large number of scientific and healthcare professionals responded to this directive in spite of many difficulties. Results from this nationwide collaboration are acclaimed internationally. When reading these reports I can see all the recorded events so clearly as if they happened just yesterday.

To write these reports, colleagues involved have collected and verified a large quantity of historical information. The reports are based on reliable data and serve as valuable references for those interested in this little known period of our history.

I represented the Ministry of Health to attend the first meeting for malaria prevention and treatment in a Beijing Restaurant on May 23 (hence Project 523), 1967, organized by the National Commission of Science and Technology and the General Logistics Department of the People's Liberation Army. This meeting was to establish a combined military and civilian program for malaria research as recommended by Zhou Enlai and approved by the State Council.

I was a permanent member of this fifteen-year project, from its start in 1967, to its termination by the State Council in 1981. I then served in the *Qinghaosu* Directional Committee until

October, 1983. I participated in regular meetings to establish research plans, follow the progress of tasks initiated by the project, and monitor the development of *qinghaosu* and its derivatives.

To continue research on *qinghaosu* and its further development, following the termination of Project 523, the Ministry of Health and the State General Administration of Medicine formed a "Directional Committee for the Research and Development of *Qinghaosu* and its Derivatives", more simply called "*Qinghaosu* Directional Committee". As the first chairman of this committee, I started the cooperation with WHO for further research on *qinghaosu.*

In those fifteen years as a member of Project 523 leading group, and first chairman of the *Qinghaosu* Directional Committee, and an official of the Science and Technology Section of the Ministry of Health responsible for supervising all medical and scientific work in China, I felt privileged to have personally worked on Project 523 and with all those involved, particularly the military personnel and field workers in villages of the Lai, Meo and Thai minority groups. We shared many good and bad times together and these events I have recorded in a book "A Collection of Photographs and Texts" by Chen Haifeng.

Although Project 523 was terminated in the early 1980s, the large number of antimalarial drugs (including chemical compounds and compounds with *qinghaosu*) developed during that period, after aiding our ally, now has a more important role in controlling malaria worldwide.

WHO recommends treating malaria with drugs in combination with *qinghaosu* (artemisinin), namely artemisinin combination therapies (ACTs). Coartem (artemether with lumefantrine) is specified as the first drug of choice by WHO, Medecins Sans Frontiers, and the World Fund. Almost all the new antimalarial compounds (ACTs) involve products from Project 523 developed in the 1970s to 1980s. It is not an exaggeration to say that a new era for a Chinese antimalarial drug has arrived. The Chinese people and Project 523 scientific workers in particular should be very proud.

Research and development of a new drug is a huge and complicated process. With inadequate scientific resources, especially a shortage of scientific equipment and trained technicians in the 1960s and 1970s, Western authorities cannot understand how the Chinese could develop *qinghaosu* to international standards. It was a collective effort of many segments of a socialist society.

Mr. Huang Shuze, the Deputy Minister of Health, stated in his 1981summary report of Project 523: "The characteristics of Project 523 included mobilizing multiple departments, involved large geographic areas, required expert planning and organization. Project 523 fully utilized each unit and department and made the best use of personnel and the available resources. The meticulous planning and tight supervision of Project 523 by the head office ensured that all tasks were carried out expeditiously, efficiently and to a high standard."

Beginning in 1975, thirty scientific research units and medical and pharmacology schools formed the "*Qinghaosu* Research Collaborative Group", to systematically work on the cultivation of *Qinghao* plants, the pharmacology and toxicology of *qinghaosu*, its chemical structure, its clinical effectiveness, and its pharmaceutical preparations with quality control for their production. It took only six years from *Qinghao* in the form of a Chinese traditional medicinal herb to being certified by the government as the drug *qinghaosu* (artemisinin). It took only fifteen years (1972 to 1987) to go through the stages from *qinghaosu* to its derivatives; from *qinghaosu* alone to combinations with *qinghaosu*; from the laboratory extraction of *qinghaosu* to large scale industrial production; to production and marketing of new antimalarial drugs containing *qinghaosu*. This degree of efficiency and speed in developing a new drug is rarely seen in China, even in Western countries with highly developed medical and pharmaceutical industries. It should again be emphasized that without a nationwide cooperative effort under Project 523, *qinghaosu* would not be the success it is.

This cooperative effort inspired teams of researchers to carry heavy back packs, climb mountains, ignore harsh weather and live in primitive conditions, as they traveled to all corners of China for several years to map the distribution and document the amount of *Qinghao* available at each location. Their work provided the material resources for building the world's first factory for industrial production of *qinghaosu* in Youyang county of Wulin mountainous district, and provided material for the production and export of *qinghaosu* antimalarials to help treat malaria in endemic areas throughout the world.

Now that China is entering a new era of a socialist market economy, the spirit and experience of Project 523 will have historic and practical significance. I hope this book will encourage medical and pharmaceutical workers in the field to pursue new drug research and development to serve the health of our people and people of the world.

Chen Haifeng.

FOREWORD

It is a pleasure and a privilege to be associated with this extraordinary account of discovery and development—not only for being involved in translating the Chinese account but also for publishing with Li Guoqiao the first article on *qinghaosu* (artemisinin) in the Western scientific literature in 1982. There is no doubt that this scientific achievement by Chinese scientists, and the comprehensive managerial oversight necessary for such a success, is a major accomplishment in the field of malaria research over the past fifty years – all the more so since this was accomplished during the decade of the Cultural Revolution, when not only were science and academic excellence in any form suppressed, but also intellectuals were unfairly and inappropriately denigrated. This present account rightly praises Chinese researchers, who searched for herbal remedies in remote areas and carried out many studies on plant extracts.

The story begins in the mid-1960s with a request from the North Vietnamese to the Chinese, for assistance in coping with a drug-resistant falciparum malaria problem in their military forces in their war against America. Mention is made of the American military in Vietnam and the South Vietnamese military having the same difficulties. I was in Vietnam around the same time as a member of the Walter Reed Army Institute medical research team, involved in the same problem for the American forces. Clearly we were on opposite sides of the same issue, although unaware of the details of the activities of each other.

Many years later I was responsible for the development of mefloquine, a new antimalarial drug discovered by the US army at the Walter Reed Institute of Research at the same time as *qinghaosu* was discovered. Although China was still a relatively closed society, in 1979 I managed to get into the country, where malaria was a problem, to study mefloquine against a control drug in a comparative trial. I anticipated this would be quinine or chloroquine, but was interested to hear about a new Chinese herbal drug with antimalarial activity. Professor Jiang Jingbo, Guangzhou Zhongshan (Sun Yatsen) University, and Li Guoqiao, Guangzhou College of Traditional Chinese Medicine (now Guangzhou University of Chinese Medicine), were my informants. I reviewed the available data on *qinghaosu* and was very impressed at the speed of parasite clearance and rapid clinical improvement in the treated patients.

Consequently we initiated a comparative study of mefloquine and *qinghaosu* and the results were dramatic. Both mefloquine and *qinghaosu* were curative but parasite clearance with the latter was much more rapid. The results were published in the British medical journal *The Lancet* (1982). The Chinese, however, had already published their studies in English in a Chinese journal in 1979. Another paper on *qinghaosu* was published in the same British journal in 1984, and we recommended that a drug combination would be the way to prevent recrudescence. Subsequently, Li Guoqiao and I collaborated on many studies over the next ten or more years, both in China and in Vietnam, where he demonstrated, among other things, that *qinghaosu* arrested the development of early ring forms of *P.falciparum* and also diminished the infectivity of gametocytes.

The outstanding properties of *qinghaosu* were very obvious, so in order to promote its wider use, I arranged in the mid-1980s for Li to take samples of the drug to Thailand to demonstrate to colleagues this valuable advance in antimalarial treatment. Unfortunately Thai malariologists were reluctant to allow studies to be done with a drug that was not approved by WHO. You will read in this monograph about WHO and the failed outcome of a five-year anticipated collaboration with Chinese experts. This book was written to correct misinformation, especially in false claims by some persons and the neglect of giving due credit to Chinese scientists for their early research and subsequent drug development.

With malaria coming under control in China with reduced opportunities for studies, clinical trials on *qinghaosu* (now called artemisinin) and derivatives were carried out in Vietnam, Cambodia, and Burma, where the scientists and doctors agreed to examine the drug without concern for approval by WHO. The clear demonstration of decreasing the incredibly high morbidity and mortality from malaria was the overriding consideration for those in the front line of patient management, when compared to the potential harm from less than ideal manufacturing conditions. "The perfect is the enemy of the good." Tran Tinh Hien, Trinh Kim Anh, and colleagues in Vietnam, just like Li Guoqiao and colleagues in China, demonstrated genuine scientific courage in advancing treatment options for malaria, such as by successfully using artemisinin suppositories in cerebral malaria. Combining mefloquine with artemisinin prevented recrudescence, and suppositories administered to children by their mothers in malaria endemic rural areas at the first sign of fever also decreased morbidity and mortality and referral to relatively inaccessible health facilities.

All these positive results had little effect on the world at large initially, and this book to some extent explains why. China, Vietnam, and Cambodia (also Burma/Myanmar) were saving many lives by using artemisinin and derivatives. Their widespread use also decreased the incidence of malaria in these countries.

Africa was the continent with the greatest need for these drugs, but it was not to benefit from their use for far too many years. China was anxious to make the drugs available outside the country, not only because the drugs were an important advance but also because China could benefit financially.

The authorities in China correctly and appropriately tried to get approval to manufacture and sell the drugs abroad, but required international approval in the form of WHO clearance. This was not forthcoming and the reasons are clearly documented here. Meanwhile the relevant authorities in Vietnam, Cambodia, and Burma decided that their populations' health was endangered and lives needed to be saved, so they used artemisinin and derivatives manufactured in Vietnam and China. African authorities were not prepared to do this and consequently many lives were unnecessarily lost in that continent in the ensuing years. In the hundreds of thousands of patients treated in China, Vietnam, and elsewhere there were few, if any, documented cases of harm resulting from a non-WHO-approved manufactured artemisinin preparation, especially when compared with the number of lives saved.

After treating over one thousand patients with the derivative artesunate, it was probably the ideal drug for development (also suggested by WHO in 1981 TDR/CHEMAL-SWG).[1] It was the most rapidly acting and effective. It could be administered orally, intramuscularly, intravenously, and even rectally, but also artemisinin suppositories were extremely effective in cerebral malaria, and easier to use in underdeveloped areas. For various reasons a derivative artemether rather than artesunate was chosen for further development by a Western company in agreement with China. There is little doubt that if, in the 1980s and early 1990s, the authorities and companies outside China had chosen artesunate and artemisinin suppositories (both already well documented as effective and safe in China and Vietnam) as the two preparations to develop, and since many patients had already been successfully treated in Vietnam, Cambodia, and Burma, most certainly many would also have benefited in Africa.

---

1   Tropical Disease Research / Chemotherapy- Scientific Working Group.

The research on combinations to prevent recrudescence and to protect artemisinin from the development of resistance also would have advanced rapidly if the collaboration between China and WHO had not broken down.

In conclusion, there is an interesting contrast to be drawn between China's achievement over the twenty or so years from the mid-1960s to the 1980s and the twenty years from the 1980s to 2006. During the first period, as this account testifies, an enormous nationwide effort was undertaken to discover and develop a drug that is a major contribution to medicine. At the end of that period, artemisinin had been shown to be very effective and safe in thousands of patients. If the drug had been made widely available immediately (preferably but not necessarily manufactured to international standards) and at low cost, to treat malaria in endemic areas throughout the world, but particularly in Africa, the disease could possibly have come under control within a decade. This was not done, and in the second twenty-year period millions of dollars from governments, foundations, and many other organizations have been poured into replicating work already done, with minimal progress made in therapy, and still malaria is not under control. In fact the latest successful combination from Li Guoqiao goes the full circle since it combines artemisinin with piperaquine (Artequick), which have been around for over thirty and twenty years respectively.

Keith Arnold, MD FACP
California

PREFACE

To meet the urgent need of our Vietnamese ally and to prepare for our own combat readiness, on May 23, 1967, in Beijing, the National Commission of Science and Technology and the General Logistics Department of the Chinese People's Liberation Army organized a meeting to coordinate research on the treatment of malaria (code-named Project 523). The aim of the meeting was to organize a multidisciplinary research team to discover and develop new drugs to prevent and treat the drug-resistant falciparum malaria that was severely affecting military operations. The team consisted of professional and educational institutions involved in medical and pharmaceutical research and manufacturing. These organizations were under the authority of various ministries, national commissions, and military affiliations, and were also from ten provinces, cities, autonomous districts, and related military districts. This was the beginning of a massive new drug research and development program.

Under unique historical conditions, Project 523 assembled a research group of over five hundred scientists from sixty of the nation's research units. With Chairman Mao's encouragement and Prime Minister Zhou Enlai's support, Project 523's national leading group guided participating scientists in achieving the important goal of keeping our troops combat ready and also to help our ally. Our leaders' encouragement was a great stimulus to motivate the team in persevering with this difficult task. After thirteen years of hard work (1967–1980), the team had developed a series of over one hundred effective antimalarial compounds and related products that included insecticides against mosquitoes and mechanical spray devices. These products were useful as first-line emergency measures for the treatment and prevention of malaria, and we had achieved the mission of military combat readiness and helping our ally. Meanwhile, from the knowledge and experience gained from, and research into, the Chinese traditional herbal medicine heritage, a new generation of antimalarials was discovered—*qinghaosu* (artemisinin) and its derivatives.

The discovery of *qinghaosu* and its derivatives is possibly the next most important discovery after quinine in the world history of antimalarial drug research. It is an important scientific contribution attesting to our country's reputation in medical and pharmaceutical research, as well as a triumphant achievement for socialism. It is also an enormous advance in the worldwide effort to control malaria that emanated from China.

For many years since 1995, after verifying and confirming the Chinese research results on *qinghaosu*-based antimalarial drugs, the World Health Organization (WHO) has repeatedly listed the Chinese manufactured artemether, artesunate, and combination artemether-lumefantrine in its "Essential Medicines List," editions 9, 11, and 12, recommending these drugs to the world. This is a confirmation of the value of new drugs discovered and developed by China. This is the first time in history for Chinese-made artemisinin-based antimalarial drugs to be entered into WHO's Essential Medicines List.[1] In January 2006, WHO declared artemisinin and derivatives to be the hope for the future worldwide eradication of malaria, and has specifically requested any country when modifying or changing its current antimalarial policy to use artemisinin-based compounds.

Goodman and Gilman's *The Pharmacological Basis of Therapeutics* is an internationally recognized, classic textbook on pharmacology. Since its first edition in 1941, a new edition appears every five years, and the latest eleventh edition was published in 2005. The textbook is read and sold worldwide, and for more than forty years has listed chloroquine as the first-choice antimalarial drug. But since the tenth edition in 2001, it has listed artemisinin and its derivatives as first-choice antimalarial drugs, instead of chloroquine. Chloroquine and its family of drugs dropped to third place on the list.[2] In 2004 the US national press announced that the committee of the Institute of Medicine (IOM) mentioned the economic problem regarding antimalarial drugs and suggested the use of combined treatment (ACT or artemisinin combination therapy) with artemisinin as soon as possible, and mentioned that this is the most important means for malaria control. *Qinghaosu* (artemisinin) is the only Chinese-made drug that is approved by the FDA to be sold in the US.

On March 14, 2002, an article in the journal *Far East Economic Review* entitled "Chinese Revolutionary Medical Discovery—*Qinghaosu* against Malaria" mentioned that at the height of the Vietnam War, the Chinese, in order to help Hanoi, had to urgently find an effective antimalarial drug. With a nationwide, large-scale, multidisciplinary and cooperative research project underway, Chinese medical scientists successfully applied their research discoveries to treat malaria patients. Since that time scientists at research centers have developed more effective *qinghaosu* (artemisinin) derivatives. Professor Richard Haynes has said that "This discovery is the greatest medical discovery of the second half of the 20th century…it is a great Chinese scientific discovery…what amazes their Western colleagues is the fact that the Chinese researchers were doing their major scientific experiments using outdated instruments that had long been abandoned by the West."[3] Given China's instability during the 1960s and 1970s and with outdated scientific facilities and equipment, Western scientists cannot understand how the Chinese could, in such a short time, have achieved such high-quality results.

With the increasing international recognition of artemisinin-based antimalarial drugs, some Chinese and overseas scholars

Zhang Jianfang

have expressed great interest in the history of *qinghaosu* (artemisinin) research by inviting articles from some research units or researchers, or requesting interviews with them. Some have already started writing the history of the discovery of *qinghaosu*. Consequently, at the present time, the history of the discovery of *qinghaosu* has become of major interest. As the main administrators of Project 532, as members of the national Project 523 leading group, its head office and local offices, and as representing the participating scientific researchers of the project, we feel obliged to explain to the world this important period in the history of *qinghaosu* research and development, in order to provide some truth-based references for those who are or will be interested in such a history.

Project 523 was completed many years ago, and there were many units involved in the *qinghaosu* research—there were at least forty-five research units providing data to the "*Qinghaosu* Research Evaluation Conference*," and the number of short-term participating units was even more. It is not feasible for this monograph to include every unit, every mission, and every detail of the work. It can record only the important research activities and processes involved, and the major steps and breakthroughs in the research program.

* * *

Note: *Qinghao* and *Huanghao (*or *Huanghuahao)* are plants of the *Artemisia* family. The effective antimalarial extract named *qinghaosu* or artemisinin does not come from the plant *Qinghao (Artemisia apiacea hance)*, but from the plant *Huanghao* or *Huanghuahao (Artemisia annua)*; therefore the name *huanghaosu* (or *huanghuahaosu*) would be more appropriate than the commonly or traditionally used name *qinghaosu*. Today this misnomer of *qinghaosu* is used interchangeably with artemisinin. Readers can find details on the plants in Chapter 2 and naming of *qinghaosu* in Chapter 5. Dihydro-artemisinin from artemisinin is the intermediate for synthesis of the derivatives artesunate, artemether, and arteether.

CHAPTER 1

Initiation of Project 523 and Its Historic Contribution to the Treatment of Malaria

Background

During the 1960s, the Indochina War escalated, eventually involving millions of soldiers and other military personnel and civilians. High rates of the disease malaria were sapping the fighting strength of the North Vietnamese army, and the old drugs were becoming much less effective because of drug resistance. The Vietnamese authorities turned to China for help. Chairman Mao Zedong and Prime Minister Zhou Enlai understood the seriousness of the malaria problem, which affected all troops in tropical areas, including their own. Mao and Zhou Enlai ordered all capable departments in China to collaborate in the urgent and important task of preventing and treating malaria by using some recently discovered drugs and drug combinations, and to discover new drugs, in order to keep their ally's troops combat-ready. A large research team was organized, comprising both military and civilian units, medical and pharmaceutical research institutes, drug manufacturing facilities, and university clinical medicine departments. The task was to develop new compounds effective against drug-resistant falciparum malaria by studying, in particular, Chinese traditional herbal remedies. This extensive and remarkably persistent and thorough effort resulted in the discovery of *qinghaosu* (artemisinin). Few people are aware of why so much emphasis and effort was put into this project, and few appreciate the magnitude of its success. This chapter tells the story of how the search for new drugs to prevent and treat malaria was begun and organized.

***

A. Malaria: The Invisible Enemy on the Battlefield

Malaria is the invisible adversary of soldiers during war. Military experts have long considered the disease a major problem affecting the fighting ability of troops during combat. Since ancient times, the armies of China and other nations throughout the world have suffered military defeats due to the ravages of malaria. In war, troops move from place to place, including malaria-endemic regions. When soldiers from non-malarious areas—with no acquired immunity—enter an endemic area, they are likely to become infected. Infected soldiers moving into a non-endemic area—if the local mosquitoes are capable vectors—can cause serious epidemics in military and civilian populations.

***

Malaria in Military History

According to ancient historical records, malaria—referred to as "miasma"—thwarted many military campaigns in China. Among the earliest references, the book *The Romance of Three Kingdoms* (as China was divided into at that time), written in the fourteenth century, recounts an event in the 300s AD describing how Zhuge Liang captured Meng Huo seven times, but it notes that many soldiers died from miasma and the military advance was hindered in the Lu River region.

The great work of Chinese history *Comprehensive Account for Aid in Government* recorded that in 745 AD (thirteenth year of the Tang Dynasty) Li Mi, a deputy minister and the governor of Jian Nan Province, led seventy thousand soldiers to attack the Tai He City of Nan Zhao Kingdom (nowadays Ta Li City in Yunnan Province). Seventy to eighty percent of the troops died of malaria and starvation, the entire army collapsed, and Li Mi was captured. Tang Dynasty poet Bai Juyi later wrote about this episode, "One heard of a river Lu in Yunnan / where miasma was dense in the month of June, / thousands of soldiers were crossing the river, / two to three, or more, in ten died."

Malaria has also affected the course of major wars throughout the twentieth century. During World War I (1914–1918), soldiers from all the nations involved died of malaria on European battlefields in the tens of thousands. In World War II (1939–1945), troops on both sides suffered from malaria, especially American and Japanese troops fighting in the Pacific War Theater. The most notorious example was in 1944 at the Indian-Burma border. Before the battle of Imphal was fully underway, sixty percent of the one hundred thousand Japanese troops developed malaria and were incapacitated before they could fight.[4]

Chinese troops had a similar experience during the liberation of southwestern China from rebel groups in the early 1950s. The malaria incidence rate was initially about 250 per 1,000 along the Yunnan border, where fighting had taken place and troops were stationed to maintain control. It took a large-scale antimalarial control effort by the military and local authorities to control the disease transmission and protect the national border. Again in 1960, malaria epidemics occurred in troops in Yunnan and many regiments lost combat strength, hindering their ability to carry out their mission.

***

The Vietnam Conflict

The US troop buildup in Vietnam began around 1964, and the Vietnamese people in the north and south fought to defend their country. Troops from both sides in the war were severely affected by malaria. Four to five times the number of American soldiers were relieved from combat duty due to malaria than were injured in combat. In 1965, fifty percent of American soldiers in Vietnam had malaria at least once. In one military action in the area of Pleiku to the border of Cambodia, twenty percent of the soldiers had malaria. In less than two months, the malaria rate in some of the regiments reached nearly one hundred percent. During the four years from 1967 to 1970, official reports declared that eight hundred thousand American soldiers were taken out of combat because of malaria, but the actual figure may have been much higher, according to remarks by the US Army head of preventive medicine. Officials of the US Army Health Department said that malaria was the number one medical problem for American soldiers in Vietnam.[4]

Similarly, the North Vietnamese troops going into South Vietnam were severely affected by malaria. Under heavy US bombing and other military action, it took some North Vietnamese troops more than a month of marching to get to the South's field of battle. When they arrived at their destination, the equivalent of only two companies in a regiment could actually fight. The rest had to be sent back to the rear for malaria treatment.[4]

***

Drug Resistance during the Vietnam Conflict

The Indochinese peninsula is in the tropics, with mountains and forests, rain, high humidity, and hot weather. Mosquitoes proliferate all year round and malaria is endemic. By the mid-1960s in Vietnam, malaria strains resistant to common antimalarial drugs, such as chloroquine, pyrimethamine, chloroguanide, and quinacrine, were common. The mortality rate from cerebral malaria, the most serious form of malignant malaria, was very

4

high and made worse by the difficulty in treating drug-resistant forms. The development of new, effective antimalarial drugs became important to the war effort on both sides of the conflict.

On the American side, the US military established a malaria research unit, increased the malaria research budget, and recruited many other research units to help develop new drugs. Experts from the Walter Reed Army Institute of Research (WRAIR) and the Navy Preventive Medicine Research Institute and specialists from universities were sent to the battlefields of Vietnam. They were to study medical and surgical problems and tropical diseases, carry out preventive and therapeutic studies, and consult on the malaria situation.

The Walter Reed Army Institute of Research in Washington, DC took the lead, collaborating with research organizations in England, France, and Australia and some large pharmaceutical companies in Europe. A large amount of money and manpower was invested in the search for chemicals to be screened as potential new antimalarial drugs. The military aimed for clinical trials on thirty new drugs per year.[4] By 1972, WRAIR had screened 214,000 chemical compounds,[5] but they had not found the ideal new antimalarial drug. This effort led, however, to the development of mefloquine. But mefloquine had side effects, and it was possibly less effective than the new antimalarial drugs resulting from our efforts, especially *qinghaosu* (artemisinin).

At the same time that the Americans began their search, the Vietnamese authorities asked China for assistance with new antimalarials. The Chinese leaders agreed to the request and thus began the mission of research for a drug to prevent and to treat drug-resistant falciparum malaria. The nationwide effort that was initiated involved medical, pharmaceutical, and technological capabilities to help Vietnam confront its American opponent on the medical front.

***

B. Initiation of Research on New Preventive and Curative Treatments

It was imperative that work begin as soon as possible on the antimalarial drug project to aid the Vietnamese ally. Within the military system, the Military Academy of Medical Science (MAMS), the Second Military Medical University in Shanghai, and the Institutes of Military Medical Science of the military districts of Guangzhou, Kunming, and Nanjing immediately started research activities.

At about the same time, in 1966, experts in the Microbiology and Epidemiology Institute of the MAMS and the Institute of Toxicology and Pharmacology suggested a review of data on the available antimalarial drugs for prevention. As a result, they identified three products, all of which were effective at preventing malaria, even against strains that were resistant to the commonly used drugs.

First, pyrimethamine and dapsone were combined in a fixed ratio and called Malaria Prevention (MP) tablet No.1. This combination had been shown effective and clinically safe in the field by researchers Ren Deli and Tian Xin from the Microbiology and Epidemiology Institute of the MAMS. This combination was effective as prophylaxis up to ten days, but to prolong the duration of action, dapsone was replaced by sulfadoxine in MP tablet No. 2. The effective duration of action of MP tablet No. 2 was two weeks.

By 1969, the Second Military Medical University in Shanghai and the Shanghai Institute of Pharmaceutical Industry (SIPI) had developed a drug with even longer protection. MP No.3 contained piperaquine and sulfadoxine and was needed only once per month. The rapid development of these three combinations of existing drugs provided the Vietnamese troops with a means to prevent infection with drug-resistant *Plasmodium falciparum*. But research for new drugs was still needed.

***

C. Project 523: A Nationwide Cooperative Project

In 1967, China was in the midst of the turmoil of the Cultural Revolution. The need for malaria drugs for Vietnam was urgent, but the political and social climate made the research extremely difficult. It was also clear that the mission could not be accomplished in a short time if only military research units were involved. The only way to complete the important task of helping our Vietnamese ally was to mobilize the resources of the entire nation, including civilian and industrial research resources. Consequently, the General Logistics Department of the Chinese People's Liberation Army (PLA) invited the cooperation of scientific research units, clinical medical units, educational units, and drug manufacturing units of various organizations. These included the National Commission of Science and Technology of the Ministry of Health, the Ministry of Chemical Industry, the Commission of Science, Technology and Industry for National Defense, the Chinese Academy of Sciences, and the leading companies in the pharmaceutical industry. All were assigned to work according to a unified plan of action.

As a first step, MAMS drafted a plan for a three-year research program. This plan was discussed at a national meeting convened by the National Commission of Science and Technology and the General Logistics Department of the PLA. Provincial, city, district, and military organizations attended.[6] The meeting was held in Beijing on May 23, 1967. Because this was an urgent military project to aid our ally, the date of the meeting was used as a code name. Thus was born "Project 523."

The MAMS plan would bring together experience and knowledge of a large number of people and organizations throughout the country. It would utilize the knowledge from Chinese traditional medicine and from Western medicine. Tasks would be assigned according to expertise, facilities, and resources, but all under a unified plan. From the plan, a multidisciplinary scientific and technical research team was formed, consisting of up to five hundred workers from more than sixty military and civilian units.

***

Project aims were to develop drugs that:

- solved the problem of drug-resistant falciparum malaria
- combined Chinese traditional and Western medicine
- emphasized the discovery of new drugs from the Chinese traditional medicine heritage

***

It was further required that the drugs developed should be:

- safe with minimal adverse effects
- highly effective with rapid onset of action
- administered infrequently, if used for prevention

***

Pharmaceutical and physical properties should include:

- resistance or protection from humidity, fungal contamination, heat, and light
- small in size and easy to transport and use [6]

***

Project 523 was begun at the height of the Cultural Revolution, when scientific research was in a state of paralysis. However, this urgent and special project to help our Vietnamese ally and to maintain military readiness was of great concern to Chairman Mao and Prime Minister Zhou Enlai. With their recognized support, and the positive spirit of the 523 Meeting, acceptance of the assigned tasks quickly spread to the different units, where scientific personnel were organized into active research teams.

Clinical research teams were formed from village "doctors" and regular doctors and sent to malaria endemic districts around China to carry out large-scale malaria preventive drug studies. The Chinese traditional medicine teams started screening old records and visiting local people in malaria-endemic areas. They went to Hainan Island in Guangdong Province, along the border of Yunnan Province, and in Jiangsu and Zhejiang Provinces, where they collected information on local malaria treatments that were secret and time-honored. The workers collected samples of Chinese medicinal herbs, some of which were tested on-site with clinical evaluations.

Another team responsible for synthetic drug research immediately began collaborating with pharmaceutical companies, manufacturers, and factories to begin investigating new chemical structures and screening for new drugs.

***

D. An Outline of the Management Organization of Project 523

Strong leadership was necessary to accomplish the project, so the "Leading Group of the National Malaria Research Team" was formed. It comprised six departments, with the National Commission of Science and Technology as the group leader, and the General Logistics Department of the PLA as deputy group leader. Other members were as follows:

- Commission of Science, Technology and Industry for National Defense
- Ministry of Health
- Ministry of Chemical Industry (MCI)
- Chinese Academy of Sciences

The head office of the leading group was located in MAMS, initially under the leadership of Major General Pen Fangfu, Deputy Director, and later under the director, Major General Qi Kairen. Bai Bingqiu served as director of the office, and Zhang Jianfang was the deputy director. The head office was overseer of Project 523 and was responsible for accomplishing the mission by organizing, assigning, and coordinating the workforce in the different districts and between units and the various specialties.

The leading group consisted of representatives from all the related departments, with the representatives managing their respective areas of responsibility. They also facilitated and coordinated the cooperation between departments so that communication was maintained with the head office.

All departments involved established a local leading team and a local office for Project 523 at the province, city, and district levels and at the military districts in Beijing, Shanghai, Guangzhou (including Hainan), Nanjing, Kunming, and Sichuan. Prominent officials headed the local leading teams and were responsible for staffing and supervising the research tasks assigned to their locality, mobilizing the local workforce to carry out the assigned task, and functioned as the relay and communication link for the smooth running of the daily work.

Four specialty collaboration groups were formed to investigate the chemical compounds present in medicinal substances used in Chinese traditional medicine for the prevention and treatment of malaria. Later, research on acupuncture, treatment for malignant malaria, immunity to malaria, and substances and devices for mosquito eradication were added to the specialty groups' responsibilities. These specialty groups helped design research projects involving cooperation between different specialties and facilitated the exchange of technical data.

***

E. The Results of Project 523

From 1967 to 1980, Project 523 completed high-quality scientific research studies on eighty-nine compounds. *Qinghaosu* (artemisinin) was the greatest accomplishment, but others were notable and received a number of awards from the Chinese government.[7] Fifteen compounds received the Chinese Science Conference Award, and twelve received the State Provincial Science Achievement Award.

Hundreds of thousands of patients in the villages suffering from malaria and other diseases were treated during the course of clinical trials. From the late 1960s to the early 1970s, large amounts of malaria prevention tablets MP No.1, MP No.2, and MP No.3 were distributed and used to help our Vietnamese ally.

Shanghai Pharmaceutical Factories Nos. 2, 11, and 14 made important contributions by modifying and improving the initial experimental products following early clinical trials. The factories manufactured several hundred tons of these antimalarial drugs to help our ally. In addition several hundred kilograms of *qinghaosu* (artemisinin) for our own Chinese troops were produced by eight manufacturing units.

Since the 1980s, twenty new products based on the discoveries of Project 523 were submitted by various units for National Invention Awards, for National Science-Technology Progress Awards, and for new drug certificates and approval for production. Thus, the products resulting from Project 523, in addition to helping our ally, also made a significant contribution to the prevention and treatment of malaria in our own country as well as having obvious economic benefits.

During the more than ten years of scientific research under Project 523, chemical compounds were developed for the acute and emergency treatment of malaria as well as for the prevention and even eradication of malaria. The research units designed and synthesized more than 10,000 chemical compounds and screened some 40,000 chemical samples to obtain nearly 1,000 possible effective substances, 38 of which underwent preclinical pharmacology and toxicology testing, and 29 of which were approved for clinical trials.

Fourteen drugs passed specialty certification and were widely used.[7] These were as follows:

1. MP No.1 (pyrimethamine and dapsone)
2. MP No.2 (pyrimethamine 12.5mg and sulfadoxine)
3. MP No.3 (piperaquine and sulfadoxine)
4. Piperaquine
5. Pyronaridine phosphate
6. Pyracrine phosphate
7. Pyracrine phosphate injection forms
8. Fansidar (pyrimethamine 25mg and sulfadoxine)
9. Hydroxypiperaquine
10. Pyrozoline
11. Naonuejia (effective for cerebral malaria)
12. Nitroquine
13. Hydroxypiperaquine phosphate
14. Diethyl amine cellulose acetate (drug coating substance)

***

Of current interest, a compound combining pyronaridine phosphate, which was developed by Shanghai Parasitology Research Institute (one of the important units in Project 523), and artesunate are under foreign development with the Medicines for Malaria Venture (MMV).

Many products of Project 523 received the National Award for Invention or the National Science-Technology Progress Award. For example, lumefantrine was awarded first place, naphthoquine phosphate was awarded second place, and both artemether and artesunate were placed third place in the National Award for Invention. The results of the extensive use of *qinghaosu* (artemisinin) received second place in the National Award in Science and Technological Advancement. Many others received awards for science and technology advancement at local departmental, provincial, city, or district level.

Cooperative teams also produced chemicals and devices for mosquito control and eradication. The patented products included compounds such as p-menthane-diol (PMD), which is extracted from the tree *Eucalyptus citriodora*, and wampee, a fruit grown in Guangxi Province. A chemical mixture, decanoate and acetoxycarvotan acetone, was developed for treating mosquito nets. Low-volume spray devices were also developed. All of these products helped solve mosquito problems for soldiers in north and south China. Some products later became commercially available to the public, especially those for mosquito control. The low-volume spray devices were attached to airplanes and used to prevent the spread of infectious diseases in Tangshan after earthquakes and have now been used extensively in agriculture and forestry for disease prevention and the eradication of harmful insects, with enormous social and economic benefits.

The treatment of malignant malaria, especially cerebral malaria, was greatly improved under Project 523. The mortality rate in 275 cases was only 7 percent through 1980, compared to 20–30 percent reported among patients treated outside China during the same period. Establishing the relationship between the *in vivo* parasite development stage and clinical symptoms of falciparum malaria, and the use of intradermal blood smears instead of bone marrow and peripheral blood smears for diagnosing cerebral malaria were important research results. These are referred to in WHO's publication *Malaria: Principles and Practice of Malariology* and also in the *Oxford Textbook of Medicine*.

Important progress was also made in malaria immunity and vector transmission. In addition, because of China's isolation from foreign research advances, our own experimental animal models and laboratory techniques had to be developed in order to study the newly discovered drugs.

CHAPTER 2

Project 523 and the Discovery of Qinghaosu (Artemisinin)

Project 523 decided to follow two pathways in the search for drugs effective against malaria (and in particular against drug-resistant falciparum malaria). One path was to screen large numbers of chemical compounds for a new active chemical, and the other, with more manpower assigned, was to concentrate on screening Chinese traditional medicines to look for an effective new product.

The Chinese herbal medicine team also took two approaches. One was to study the traditional remedies passed on usually in secret from generation to generation, undertaken by many small research groups, which were formed in Beijing, Guangdong, Sichuan, Yunnan, and Jiangsu Districts. The members of these groups interviewed the local population and collected herbal medicine samples claimed to be effective for treating fevers presumed to be malaria. Crude laboratory extracts from these samples were screened for antimalarial activity and subjected to drug safety testing before being evaluated in clinical trials. When one group found a promising substance, all the other groups joined together to speed up the development process by each group concentrating on a particular aspect of the development program. The other approach was to review ancient and modern traditional pharmaceutical written records. Based on this survey of documents and using the criteria of frequency of citation in the literature, ten research targets were identified, one of which was the Chinese herb *Radix dichroae* and another *Qinghao*.

The initial effort was directed at modifying the chemical β-dichroine in *Radix dichroae* because, although it was known to have antimalarial activity, it caused severe vomiting. This modification project was carried out simultaneously in Beijing, Shanghai, and Sichuan.

In Beijing the work was carried out collaboratively by Deng Rongxian of the Beijing Pharmacology Institute of the Academy of Military Medical Science (abbreviated as Beijing Institute of Materia Medica) and Jiang Yunzhen of Beijing Pharmaceutical Factory (which was a merger between the original Beijing Pharmaceutical Industry Research Institute and Beijing Pharmaceutical Factory during the Cultural Revolution). This factory was where the actual research was done. First β-dichroine of the Chinese herb *Radix dichroae* was chemically synthesized, and then its molecular structure was modified to create a series of derivatives. Of these the most effective chemical derivative was coded as "7002." It underwent pharmaco-toxicology studies in the Beijing Institute of Materia Medica and then clinical trials in Hainan Island from 1970 to1973. The results demonstrated that compound 7002 was more effective and had less toxic, adverse side effects than β-dichroine. In a combination form it was even more effective, but vomiting as a side effect was more frequent. Because of this high vomiting rate and the emergence of other Chinese herbs with better antimalarial activity, research on this new substance was terminated.

The Shanghai Institute of Materia Medica of the Chinese Academy of Sciences studied the chemical structure of β-dichroine and also designed a series of derivatives. Preclinical data and good clinical results were obtained with a derivative called "56," but because it was not as effective as other new drugs that were under development at the same time, it was not widely investigated.

The Sichuan Institute of Chinese Materia Medica was also responsible for various research projects on the Chinese herb *Radix dichroae* and its active component β-dichroine. This institute supplied large amounts of β-dichroine to other units for their research. The Sichuan Medical College and Chengdu University of Traditional Chinese Medicine performed clinical studies and further research to decrease the vomiting side effect of β-dichroine.

Although research attempts at modifying the chemical structure of *Radix dichroae* and β-dichroine did not bring about a better and newer antimalarial drug, it did reflect on the high academic and technological standard of the Chinese Chemical Pharmaceutical researchers and industry. Also the new chemical structures led to two new non-antimalarial drugs. One was alexipharmic, a new antidote for neurotoxic organophosphorus pesticide poisoning, and another was for treating cardiac arrhythmias.

From 1967 to 1969, as part of the three-year plan of Project 523, the Chinese traditional medicine and pharmacology collaboration group in Beijing District consisted of personnel from the Beijing Institute of Materia Medica and the Military Academy of Medical Science. This group worked with scientific and technological personnel in Yunnan, Guangdong, and Jiangxu to form several investigative groups.

These groups went into the local population in their own district to collect medicinal herb samples to test extracts for drug effectiveness and to perform preliminary clinical trials. From the work in Guangdong Province, the medicinal plant *Artabotrys uncinatus* appeared to be a most promising lead.

In 1969 the Beijing Institute of Materia Medica collaborated with Sun Yat-Sen Medical College, Sun Yat-Sen University, and the South China Botanical Institute of the Chinese Academy of Sciences (CAS) (all three in Guangdong Province), to study the antimalarial effects of this medicinal plant, *Artabotrys uncinatus*. One of the main investigators was Yu Dequan of the Beijing

Institute of Materia Medicina (now a member of the Chinese Academy of Engineering, CAE). The antimalarial effect of *Artabotrys uncinatus* was positive, and it was one of the priority research projects for the "Medicinal Herbs Specialist Group" at the time. The active element in *Artabotrus uncinatus* was identified and named *yingzhaosu* A, which had potent antimalarial effects. The Beijing Institute of Materia Medica collaborated with the Chemistry Department of Sun Yat-Sen University and identified the chemical planar structure of *yingzhaosu* as a fat-soluble peroxide containing no nitrogen, a new class of chemical compound. But the amount of *yingzhaosu* A contained in the plant *Artabotrus uncinatus* was very low and difficult to extract in amounts large enough for use. From 1974 to 1977, the Chemistry Department of Sun Yat-Sen University synthesized compounds similar to *yingzhaosu* A in more simplified forms. The presence of a peroxide group in the chemical structure of *yingzhao* was of help in identifying the chemical structure of *qinghaosu* later.

In summary the specialist group on "Chinese Traditional Herbal Medicine" spent several years screening medicinal plants and herbal remedies and found ten promising antimalarial medicinal products. These included *Artabotrys uncinatus*, *Herba agrimoniae*, *Artemisia*, and *Polyalthia nemoralis*. The group also identified many active compounds from four of the Chinese medicinal herbs. *Qinghaosu* (artemisinin), which is extracted from the herbal plant *Artemisia annua L.*, was the most important discovery, a discovery that is at last recognized worldwide.

***

A. The Plant Artemisia Annua L. or "Qinghao"

The Chinese people and medical professionals have had a long history of struggle against malaria. The term "hao" and "malaria" already existed in the inscriptions on bones and tortoise shells in the ancient Ying Dynasty. In the manuscript "Prescriptions for 52 Diseases," excavated from the two-thousand-year-old Mawangdui Han tomb in Hunan Province, there was a record of using *Qinghao* (*Artemisia annua L.*) to treat diseases. In the book *Shen Nong Ben Cao Jing*, in the second to third century AD, the name "*Caohao*" was used instead of *Qinghao*. The earliest record of using *Qinghao* for treating malaria is found in "Zhou Hou Bei Ji Fang," a handbook containing regimens for emergency treatment written by the medical and drug professional and alchemist Ge Hong in 340 AD.[8] The book *Ben Cao Gang Mu* (*Compendium of Materia Medica*), by Ming Dynasty medical and drug professional Li Shizhen, and other ancient medical books also recorded using *Qinghao* or combinations with *Qinghao* for malaria treatment. In some malaria endemic areas, the local population still uses many Chinese medicinal herbs for malaria prevention and treatment, as well as for controlling mosquitoes. Recently the medical and pharmaceutical journals from Jiangxu, Hunan, Guangxi, and Sichuan Provinces have published reports on the clinical use of *Qinghao* in treating malaria. In the general population, *Qinghao* has been used in various forms for treating malaria, such as by crushing the plant into a juice, by boiling it in water and using the extract, or as powder from the dried herb. The Project 523 meeting in 1967 inaugurated a search for new antimalarial drugs that may be found within traditional Chinese medicinal practices. With this directive and after screening traditional medical records, as already mentioned, ten antimalarial Chinese medicinal herbs were identified as priority targets for research, with one of them being the herbal plant *Qinghao* or *Artemesia annua L.*[6]

Qinghao, also called "smelly *hao*," "fragrant *hao*," and "bitter *hao*" by the general population, belongs to the plant family *Compositae* that grows widely in south and north China. Two main species of *Qinghao* are used as Chinese traditional medicinal herbal remedies—namely, *Huanghuahao* (with yellow flowers) *Artemisia annua* L., and *Qinghao* or *Artemisia apiacea hance*. Research showed that only *Huanghuahao*, *Artemisia annua L.* and its mutation with large flower heads contained the active antimalarial substance.

The Project 523 Chinese medicinal herb research group in Nanjing District learned from the local population that in 1969 rural doctors and villagers in the Gaoyou County of Jiangsu Province used the locally grown *Qinghao* for malaria prevention and treatment *en masse* in the general population and obtained good results. There was even a common local expression: "Don't worry when you get malaria—just take *Qinghao* with sugar."

Even earlier, in 1958, people in the Gaoyou County of Jiangsu Province already were using *Qinghao* to treat malaria. The Jiaoshan production brigade of the Longben community in this county initially had a high incidence of malaria. But this quickly changed from 1969, when *en masse* malaria prevention and treatment with *Qinghao* was initiated. At the same time, with the local "*en masse* antimalaria movement," local doctors observed 184 malaria patients treated with *Qinghao* from 1969 to 1972 and found the efficacy rate above 80 percent. In the summer of 1972, the Hehe production brigade of the Cheluo community had the entire community taking quick-boiled *Qinghao*, effectively controlling a malaria epidemic.[9]

In 1972 the Gaoyou County (of Jiangxu Province) formed a "Validating Group of *Qinghao* in Malaria Prevention and Treatment," consisting of the local Department of Health, the Health and Quarantine Station, and the Cheluo District Hospital. This group comprised 101 persons from the professional teams and the healers of the population at large, forming eight clinical observation groups working in five experimental areas. Each clinical observation group consisted of personnel from the local Department of Health, doctors of Western medicine, doctors of traditional Chinese medicine, and village doctors. The *Qinghao* Malaria Prevention and Treatment Testing and Certifying Group also organized training classes for village doctors, clinical doctors, and laboratory technicians. In the provincial *en masse* malaria prevention and treatment movement, these eight clinical observation groups tested the antimalarial effects of *Qinghao* by using it in many different forms, such as fresh juice, a quick-boiled form, a decoction, pills and tablets, fresh versus dried *Qinghao*, young thin *Qinghao* branches and leaves versus thick branches and leaves, long versus short duration of boiling *Qinghao*, and large versus small doses of *Qinghao*. They also improved the timing and the frequency of drug administration. In an analysis of the data, a summary of the results of several years of study showed that, of the six forms of *Qinghao* preparations, the extract as a compressed powder in the form of a tablet had the best results. It was a high-quality preparation that resulted in a shortened duration of fever and parasitemia time, high cure rate, and a low recrudescent rate.[9]

From June to October 1976, in the twenty-six observation sites of the five experimental districts of Jiangxu Province, *Qinghao* was given to 240,000 people for malaria prevention. Compared to the same period in 1975, the malaria incidence in 1976 was reduced 50 percent. *Qinghao* was used to treat 201 symptomatic vivax malaria patients with a cure rate of 89 percent.[9] In this period, at the request of the Ministry of Health, the Institute of Chinese Traditional Medicine went twice to Gaoyou County of Jiangxu Province to investigate these results. The Institute confirmed the effectiveness of *Qinghao* in treating malaria and consequently supplied important data for the decision to be made to further develop *Qinghao* and *qinghaosu*.

***

B. Further Development of Qinghaosu

As previously mentioned, several groups were assigned to work on Project 523 and to approach the problem in different ways; but inevitably there would be some overlap of activities.

In 1970 scientist Gu Guoming of the Military Academy of Medical Science collaborated with scientists Yu Yagang and Tu Youyou from the Beijing Institute of Materia Medica. They reviewed ancient and current medical records for data on the use and effectiveness of Chinese herbal remedies on malaria, as did other workers already described. Ethanol extracts were made from over one hundred frequently cited remedies, and these were tested in a rodent model of malaria at the Military Academy of Medical Science.

Among these extracts was *qinghaosu*, with a variable 60–80 percent positive result when tested in a rodent malaria model using *Plasmodium berghei*.[10] In 1971 Gu Guoming left the project and returned to his unit for other duties.[11] Ning Dianxi, from the Instituteof Microbiology and Epidemiology of the Military Academy of Medical Science, took over from Gu Guoming and was sent to the Beijing Institute of Chinese Materia Medica to help set up a mouse animal experimental model for *Plasmodium berghei* malaria. From this point on, *qinghaosu* research was carried out by the Beijing Institute of Chinese Materia Medica.

In the latter half of 1971, researchers in the Beijing Institute of Materia Medica uncovered a hint from a comment in the medical handbook "Zhou Hou Bei Ji Fang," written by Ge Hong in 340 AD. This said, "Soak one bundle of *Qinghao* in two liters of water, drink the juice obtained from compression of the soaked *Qinghao* to stop malaria." They thought that the high temperature they had used in preparing *Qinghao* might have destroyed the active ingredient in the plant, thus affecting its antimalarial properties. They then replaced the ethanol extraction method by an ether extraction method, which had a lower boiling point. The ether-extracted *Qinghao* product on testing had a success rate of 100 percent in the *Plasmodium berghei* mouse malaria model.

March 1972 in Nanjing, the national 523 Project head office called a meeting with two specialty groups namely, the chemical synthetic drugs group and the Chinese medicinal herbal group. At the meeting with the Chinese medicinal herbal group, Tu Youyou, representing the Beijing Institute of Chinese Materia Medica, reported the 100 percent recent success rate against *Plasmodium berghei* malaria in mice with the initial crude extract (*qinghaosu*). Other units of the Chinese medicinal herbal group reported on more than ten Chinese herbs and plants, such as *Artabotrys uncinatus* (*yingzhao*), *Herba agrimoniae*, and *Polyalthia nemoralis*, some of which had an antimalarial success rate of 80–90 percent. The excellent results of the *Qinghao* extract attracted the attention of the Project 523 head office and the specialty groups. The meeting suggested that the Beijing Institute of Chinese Materia Medica speed up further experimental research on the methodology for *Qinghao* extraction, drug effectiveness, and drug safety. The meeting suggested further work on the extraction and identification of the active component in *Qinghao*, and to simultaneously carry out clinical studies to confirm the effectiveness of the initial crude extract.[12]

With the support and liaison efforts of the Project 523 head office, the Beijing Institute of Chinese Materia Medica carried out further experiments and singled out the neutral portion of the ether extract as the active element, identified as code No.91. After toxicology studies in animals and a small number of tests in healthy volunteers, resulting in no apparent toxic side effects, clinical trials were carried out.

These clinical trials with the neutral portion of the ether extract were performed in the Changjiang area (of Hainan Island) from August to October of 1972, and in Hospital No.302 of the Chinese People's Liberation Army in Beijing. There were thirty patients in total. The twenty-one Hainan patients were non-indigenous islanders, eleven with vivax malaria, nine with falciparum malaria, and one with mixed vivax and falciparum malaria. The nine Beijing patients had vivax malaria. Three dosages were tested, and all were effective against malaria. The large dosage group (3 grams, four times daily for 3 days, for a total of 36 grams) gave the best results. The average temperature normalization time was nineteen hours for vivax malaria and thirty-six hours for falciparum malaria in non-indigenous Hainan patients, but malaria parasites reappeared after a short period. (Editor's note: the original clinical trial summary stated "in nine falciparum malaria cases tested, seven were effective, two were ineffective."[13, 22])

On account of the large physical size of the tablet formulation, it was not possible to test a larger dosage, but the parasite reappearance rate was lower in the high dosage group than the low dosage group. There were no serious side effects, but a few patients reported vomiting or diarrhea. No abnormal liver or kidney function tests were found. The two Beijing patients with mild pretreatment elevation of amino-transferases, which continued to rise after taking drug, normalized in two to three weeks. See summary tables 1 and 2.[14]

The research progress made on *Qinghao* by the Beijing Institute of Chinese Materia Medica from 1972 to 1973 was very significant for the later research and the discovery of *qinghaosu* (artemisinin). The crude ether extract in the latter half of 1971 had a rodent malaria success rate of 100 percent in mice, and the clinical trials of the neutral portion of the crude ether extract in 1972 gave good results. Both confirmed the antimalarial effect of the Chinese medicinal herb *Qinghao* in laboratory animals and clinical patients. This important step forward for *Qinghao* research motivated later *qinghaosu* research.

The good results of the clinical trials of the crude extract in Hainan and Beijing brought hope for developing antimalarial Chinese medicinal products. In 1972 simultaneously with the clinical trials, the Project 523 head office requested the Beijing Institute of Chinese Materia Medica to identify and separate out the active element in the crude extract. By the end of 1972, the institute had identified and separated out several active constituents, one of which had an antimalarial effect and was named "*qinghaosu* II."

The 1973 research plan of the Project 523 head office requested rapid completion of preclinical animal toxicology testing of *qinghaosu* II so that clinical trials could begin. In the animal toxicology study in the Beijing Institute of Chinese Materia Medica, *qinghaosu* II had cardiac toxicity in tested animals. Therefore opinions differed as to whether *qinghaosu* II clinical trials should start. Professor Jing Houde and other scientists from the Pharmacology Laboratory felt that clinical trials should not start because of the cardiac toxicity; others suggested testing it on research subjects first, and if no cardiac toxicity was detected then to start clinical trials immediately. On this issue the leaders and specialists of the Institute had many discussions that were reported to Project 523 head office. The head office recommended a thorough review of the issue. Finally, *qinghaosu* II was tested on three researchers and showed no apparent toxicity. Leaders of the Institute therefore agreed to start clinical trials of *qinghaosu* II.

In September and October 1973, traditional medical doctor Li Chuanjie of the Acupuncture Institute of the Academy of Chinese Traditional Medicine headed a clinical research team that again went to Hainan, Changjiang area to start a clinical trial. Shi Linrong from Project 523 head office and Deputy Director Wang Lianzhu of the Shanghai 523 office went to Hainan to inspect the fieldwork, and to the Shilu region of Changjiang area to observe the *qinghaosu* II clinical trial.

It was understood that the clinical trial in the Changjiang area of Hainan would have non-indigenous patients with either vivax or falciparum malaria. Because of an unsatisfactory antimalarial effect and the appearance of cardiac toxicity, the *qinghaosu* II clinical trial was terminated after eight cases. The Project 523 head office did not receive a report of this clinical trial, but later read about it in the publication of the Beijing Institute of Chinese Materia Medica. It was a short report:

"From September to October 1973 a clinical trial with *qinghaosu* (note: i.e.*qinghaosu* II) was carried out in the non-indigenous population, with vivax and falciparum malaria, in the Changjiang area of Hainan Island. A total of eight cases were studied, three with vivax, and five with falciparum malaria. In the vivax malaria group a total dose of 3–3.5 grams of the drug was given in a capsule formulation. The average temperature normalization time was 30 hours, and at follow-up three weeks later two cases were cured while one case had reappearance of parasites on day 13. In the falciparum malaria group, one patient with a parasitemia above $70,000/mm^3$ given 4.5 grams in tablet form normalized the temperature in 37 hours and cleared the parasitemia in 65 hours, but parasites reappeared on day six. Two cases were terminated because of cardiac toxicity as shown by premature ventricular contractions. Another patient, with falciparum malaria for the first time, had a parasitemia of 30,000/mm^3 and was given 3 grams orally. The temperature normalized in 32 hours and the parasites disappeared but reappeared one day after taking the drug, associated with a rise in temperature. Two patients did not respond to the drug and were considered treatment failures."[15]

The results of the *qinghaosu* II clinical trial in Changjiang were seriously examined because the treatment effectiveness and toxicity were very different from later results with *huanghuahaosu* in Shandong and Yunnan Provinces. (Before the 1978 National *Qinghaosu* Evaluation Conference, each locality independently named their extraction product. It was *"huanghuahaosu"* in Shandong Province, *"huanghaosu"* in Yunnan Province, and *"qinghaosu"* in Beijing.)

From 1973 to 1974, the *qinghaosu* II research program in the Beijing Institute of Chinese Materia Medica experienced some research difficulties, and progress was slow. But encouraging news did come from Project 523 units in Shandong and Yunnan Provinces, and therefore confidence was restored at the Project 523 head office to continue *Qinghao* research.

***

C. *Huanghuahao as the Hope for the Future*

After attending the 523 traditional Chinese medicinal specialty group meeting in Nanjing, the representative of Shandong Institute of Parasitic Diseases returned to Shandong, and based on the experience of the Beijing Institute of Chinese Materia Medica, he used local *Huanghuahao* extract to test on rodent malaria in mice. The results of this experiment were reported to the national Project 523 head office by mail on October 21, 1972, and were consistent with the Academy of Traditional Chinese Medicine's results of *Qinghao* extract on rodent malaria.[16] In 1973 the Shandong Institute of Parasitic Diseases collaborated with the Shandong Institute of Chinese Materia Medica in a clinical trial with a crude ether extract, "*Huang* No.1". In thirty vivax malaria patients in Juye County, ten were given a total of six capsules of "*Huang* No.1" (each with 17.1 grams of crude drug, over three days, two capsules per day). A successful parasite killing effect was obtained with a rapid control of clinical symptoms, and this effectiveness exceeded the standard three-day chloroquine regimen. Another ten patients received six capsules as a single-dose, and another ten patients two capsules twice in twenty-four hours. In these latter two groups, parasitemia either did not clear, or if it did, parasites reappeared within a few days. Side effects were mild, with self-limiting vomiting and diarrhea occurring in a few patients, mostly in the three-day dosage group. These results indicated that *Huang* No.1 was a rapid acting drug with a short duration of action.[17]

In November 1973, the Shandong Institute of Chinese Materia Medica separated the active monomer in the local *Huanghuahao* and named it "*huanghuahaosu*." Its initially established melting point was 149–151°C. The data were sent to the national Project 523 head office.

At the time that the Beijing Institute of Chinese Materia Medica was having difficulties with "*qinghaosu* II," the encouraging experimental laboratory and clinical results from

Shandong Province were highly welcomed by the national Project 523 head office. The national Project 523 head office sent representatives to Shandong Institute of Chinese Materia Medica with a member of the Beijing Institute of Chinese Materia Medica, to personally observe the *"huanghuahaosu"* research work, to exchange experiences and to promote *Qinghao* research.

In November 1973, Shi Linrong, from the national Project 523 head office, and research worker Meng Guangrong, of the Beijing Institute of Chinese Materia Medica, went to Shandong Province. Discussions were held with Wei Zhenxing of the Shandong Institute of Chinese Materia Medica on the topic of *qinghaosu* cardiac toxicity and the temperature for the extraction procedure. Wei Zhenxing showed them the extraction methodology and the data on the pharmacology testing. Their *Huanghuahao* monomer had not shown cardiac toxicity, and the problem of destruction of the active component when the temperature of the extraction process exceeded 60°C did not happen with their extraction technique. The trip to Shandong was beneficial to the research of both parties.

In late 1972, Kunming District 532 office director Fu Liangshu heard about the *Qinghao* research by the Beijing Institute of Chinese Materia Medica when attending the annual meeting of 523 district office directors held in Beijing. Director Fu Liangshu visited the Beijing Institute of Chinese Materia Medica after the meeting and saw the *Qinghao* extract there in the form of a black material. When he returned to Kunming, he informed the 523 team of the Yunnan Institute of Materia Medica of the Beijing Institute's findings and suggested that the local team widely screen their local plants of the *Artemisia* family since Yunnan had the advantage of being rich in plant resources.

In early 1973, Luo Zeyuan from the Yunnan Institute of Materia Medica visited Yunnan University. She found "bitter *hao*" grew in abundance in the university campus. She brought some samples back to the Institute's laboratory and tried extraction by boiling with four different organic solvents: petroleum ether, ether, ethyl acetate, and methanol. By April 1973, she was able to isolate the active monomers from the ether extract. Because the plant was not yet scientifically classified at that time, the Institute temporarily named its active monomers "*Kuhao* crystal III" ("bitter *hao* crystals III," later called "*huanghaosu*"), using the popular local name "*Kuhao*" (or bitter *hao*) for the plant. In the *Plasmodium berghei* malaria suppression test performed by researcher Huang Heng of the Pharmacology Laboratory, the parasites disappeared very quickly. By October 1973, the preliminary pharmacology and toxicology studies of "*huanghaosu*" were completed. With large- and small-size animal toxicology testing, there were no adverse effects detected on the heart, liver, and kidneys.[18]

To scientifically classify the plant, they sought the help of botany professor Wu Zhenyi from the Kunming Institute of Botany, Chinese Academy of Sciences. The plant was identified as *Artemisia annua L. f. macrocephala Pamp*, a variant with a large yellow flower head and a simplified name of "large head *Huanghuahao*."

In April 1973, when the successful extraction of *Huanghuahao* crystals had been accomplished, it was found that the amount of available local plant material was very small. Yunnan Institute of Materia Medica therefore bought *Huanghuahao* plants from Chongqing City. They found the 0.2 percent concentration of "*huanghuahaosu*" in the Chongqing plants to be ten times higher than the 0.02 percent concentration of the local plants. Further investigation revealed Youyang area in Sichuan Province as the origin of the purchased *Huanghuahao* plants. They sent institute members to Sichuan Youyang area many times to buy *Huanghuahao* directly and in person. Tests consistently showed a higher "*huanghuahaosu*" concentration of 0.2–0.3 percent in the plants from Youyang.[19] Thus even before the systematic investigation on a *Qinghao* resource started, this Yunnan research group had already proved Youyang area to be an excellent district for growing high-quality *Huanghao* plants, and they provided information on the origin of high-quality *Huanghuahao* plants for future *qinghaosu* research. Their findings were also a contributing factor later in choosing Youyang area as the primary site for growing *Qinghao* plants and producing *qinghaosu*.

In the autumn of 1973, Zhou Keding of the national Project 523 head office, and researcher Zhang Yanzhen of the Beijing Institute of Chinese Materia Medica went to the Yunnan Institute of Materia Medica to observe the progress of the *Huanghuahao* research and to learn about their advances in extraction techniques. The researchers of the Yunnan Institute of Materia Medica presented their research and techniques without reservation. The visitors returned to Beijing with some *"huanghaosu"* as reference samples. This event reflected the excellent attitude and spirit of information exchange between research units.

The contribution of the Yunnan Institute of Materia Medica in *qinghaosu* research did not stop at successfully extracting *"huanghaosu"* from local plants and in identifying the location of high-quality *Huanghuahao* plants. It also provided *"huanghaosu"* crystals to other teams in the early days of the research, demonstrating the cooperative nature of the participants in Project 523.

In early 1974, Zhan Eryi and Luo Zeyuan of the Yunnan Institute of Materia Medica developed an improved method for *huanghaosu* extraction called the "petroleum solvent method." This entailed an initial extraction with a petroleum solvent, concentration of *huanghaosu* extract, and recrystalization with 50 percent ethanol, giving purified *huanghaosu*. This method provided enough *huanghaosu* for animal pharmacology studies and clinical trials and therefore speeded up the research activities.

Healthy competition between research units demonstrated the efforts being expended to develop *qinghaosu*. The Yunnan

Institute of Materia Medica in April 1973 was the first to extract crystals of *huanghaosu* with potent antimalarial activity, followed in November 1973 by the Shandong Institute of Materia Medica. In the clinical trials development phase, the Shandong Institute was ahead of the Yunnan Institute. These two institutes carried out their independent research studies at the same time that the Beijing Institute of Chinese Materia Medica was having difficulties with *qinghaosu* II. The end result of all this, however, was to encourage and give confidence to the Project 523 head office that *qinghaosu* should be the main research target.

***

D. Beijing Meeting of Shandong, Yunnan, and Beijing Project 523 Local Teams

Based on the progress of *Qinghao* research in Beijing, Shandong, and Yunnan, the national Project 523 head office, on January 10, 1974, organized a meeting of all district office leaders of Project 523. The purpose of the meeting was to decide whether to focus on *Qinghao* as the target for the 1974 Antimalarial Chinese Medicinal Herbs Research Project. The head office felt the need to call all local research teams to present their work, to summarize their findings, to exchange experience with other teams, and then to establish a unified research plan. The assignment of specific tasks to appropriate local teams had the purpose of speeding up *qinghaosu* research.[20]

On February 28, 1974, a "*Qinghaosu* research meeting" was held in the Beijing Institute of Chinese Materia Medica. Attending were many scientists, including those from the Beijing Institute of Chinese Materia Medica, Shandong Institute of Chinese Traditional Medicine, Shandong Institute of Parasitic Diseases, and Yunnan Institute of Materia Medica. Attendees included Director Liu Jingming of the Beijing Institute of Chinese Materia Medica, Wei Zhenxing of the Shandong

Institute of Chinese Traditional Medicine, Chief Scientist Liang Juzhong of the Yunnan Institute of Materia Medica, and Deputy Director Tang of the Shandong Institute of Parasitic Diseases. The manager and leader of the national Project 523 head office also attended the meeting and were responsible for preparing a summary report of the meeting.

Based on the research results of *Qinghao/Huanghao* of various local research teams, the meeting discussed the 1974 research plan, identifying the year's research targets and assignments. The following is an excerpt of the meeting summary:

1. Identify the chemical structure of the active component. In 1973 all three teams had independently extracted an active component in their locality. These extracts could be the same substance. The meeting suggests that the Beijing Institute of Chinese Materia Medica continue this work with the Shanghai team to identify the chemical structures of these extracts as soon as possible.
2. Research to improve extraction techniques, to identify the component for medical use, to optimize the harvesting season, and to investigate the geographic distribution of the most productive plant source. These activities will be carried out by all three teams according to their local situation.
3. Preclinical pharmacology testing: The Yunnan Institute of Materia Medica continues its research on preclinical pharmacology studies on *Qinghao* crystals, and the Beijing Institute of Chinese Materia Medica continues to work on its possible cardiac adverse effects.
4. Clinical trials: Completion of clinical trials of *Qinghao* crystals on 150 to 200 patients (50 falciparum malaria cases and 100 to 150 vivax

malaria cases). The studies will be carried out by the Shandong Institute of Parasitic Diseases and Yunnan Institute of Malaria Prevention and Treatment, with additional researchers sent from the Academy of Chinese Traditional Medicine. To ensure the availability of active crystals for the clinical trial, the Shandong Institute of Materia Medica will provide material for 150 patients, the Yunnan Institute of Materia Medica will provide for 30 patients, and the Academy of Chinese Traditional Medicine for 50 patients.

5. A summary of the research on the pharmaceutical unpurified, relatively crude preparation of *qing-haosu*.[21]

***

Following instructions issued prior to the meeting, the Shandong Institute of Materia Medica and Yunnan Institute of Materia Medica brought their own *huanghuasu* (*huangsu*) extracts for the Beijing Institute of Materia Medica.[22] In 1973 the Beijing Institute of Materia Medica and the Institute of Organic Chemistry of Shanghai Academy of Sciences collaborated in establishing the chemical structure of *qinghaosu*. Through the liaison of the national Project 523 head office, the Shandong Institute of Materia Medica had supplied more than 10 grams of *huanghuahaosu*.[22]

Following the assignments of the Beijing meeting to the three local teams, the Shandong Institute of Materia Medica and Yunnan Institute of Materia Medica extracted *huanghuasu* (*huangsu*) and carried out local clinical trials in Shandong's Juye County, and in Yunnan's Yun County. In 1974 the Beijing Institute of Materia Medica was not yet able to fulfill its assignment of extracting and supplying *Qinghao* crystals.

***

In Shandong Province, the Shandong Institute of Materia Medica, the Institute of Parasitic Diseases, and Qinghai Medical College formed a task force headed by Director Zhu Hai of the Shandong Institute of Chinese Materia Medica and researcher Li Guiping of the Institute of Parasitic Diseases. The task force organized several *huanghuahaosu* specialty collaborative groups to work on the chemistry, pharmacology, drug formulations, analysis, and clinical trials of *huanghuasu*. In May 1974, they completed a clinical trial on twenty-six vivax malaria cases in Juye County. The crude preparation was used in seven of these twenty-six cases. The regimen was a three-day treatment course with a total dose of 0.6–1.2 grams. The average time for parasite clearance was 72 hours. In most cases, parasites were not detectable in the blood 48 hours after drug administration. The clinical symptoms were controlled after the first dose. No cardiac side effects (judged on heart sounds and rhythm) were noted. But the recurrence rates were high. In the two-month follow-up, symptoms and parasites reappeared in four of the five patients. The following is the abstract of that study, entitled "Preliminary observation of vivax malaria patients treated with *huanghuahaosu* and the simple pharmaceutical preparations of an acetone extract of *Huanghuahao*."

"1. In early May and early October 1974 a clinical trial on treating vivax malaria with *huanghuahaosu* and a simple pharmacologic preparation of *Huanghuahao* was carried out in the Zhuzhuang production brigade of the Chengguan Community of Juye County. There were 26 cases, divided into four groups. Only one case in the crude pharmaceutical preparation treated group had symptoms recurring twice; the other three groups had either total control of symptoms or they only recurred once. The recurrence of symptoms usually occurred within 2–10 hours after drug administration, far earlier than the expected symptoms recurrence time. This timing coupled with the change in level of parasitemia suggested the episode was likely a'drug fever' secondary to massive parasite destruction by drug.

2. In all study groups parasites disappeared quickly after drug administration. The parasite clearance time was within 72 hours except for one patient in the low dose *huanghuahaosu* group who had a parasite clearance time of 96 hours. Twenty-four hours after drug administration there were obvious changes in the peripheral blood parasite concentration and the morphology of the parasites. There was parasite maturation arrest at the large trophozoite stage, with smudged nuclei and cytoplasm with coarse dark pigment granules. Parasite concentration was significantly lowered and became negative in some patients. At 48 hours parasites could not be detected in most cases or were present as scant morphologically altered forms. In a few cases, parasite clearance time was 96 hours.

3. In all study groups, including two pregnant patients, there were no apparent side effects and no change in heart sounds or rhythm.

4. Follow-up was two months for all study groups. In both the *huanghuahaosu* 0.2 gram times 3-day group and 0.4 gram times 3-day group, reappearance of parasites and symptoms occurred in four of five patients. But recurrence occurred early, within 15 days in some cases. In the *huanghuahaosu* 0.4 gram plus 2 tablets of Malaria Prevention No.2 group, four in nine patients recrudesced within two months which was significantly less than in the *huanghuahaosu* alone treated groups where recurrence occurred later, about one month or more than one month. In the group treated with the crude acetone extract

preparation, two of seven patients recrudesced within 15 days and the drug effectiveness was similar to *huanghuahaosu*.

5. This study indicated that *huanghuasu* is a relatively fast acting antimalarial drug. Its rapid control of clinical symptoms, massive destruction of parasites, and no apparent side effects seems to give it great value in the emergency treatment of malaria. But its action was not curative with recurrence occurring relatively shortly after drug cessation and with a relatively high rate. In the 2 *huanghuahaosu* alone groups, either the 0.2 gram times 3-day or the 0.4 gram times 3-day group, the 0.4 gram group had a faster immediate effect, but there was no difference in recurrence rates between the two groups. To effectively control the recurrence with *huanghuahaosu* alone using a higher dose was not easy and combination with other antimalarial drugs should be considered. In this study *huanghuahaosu* 0.4 gram plus 2 tablets of Malaria Prevention No.2 had maintained an immediate curative effect with a lower recurrence rate. Clinical studies in combination with primaquine should be considered. The crude pharmaceutical preparation of an acetone extract maintained the therapeutic effectiveness of *huanghuahaosu*, but combination with other antimalarial drugs should also be considered."[23]

***

The foregoing results indicated that the *huanghuahaosu* extracted by the Shandong Institute of Materia Medica had good short-term clinical effects against vivax malaria in northern China. Because there was no falciparum malaria in Shandong Province, the effectiveness of the drug against this form of the disease, and especially drug-resistant falciparum malaria, was unclear as yet.

In the autumn of 1974, the Yunnan clinical team carried out a clinical trial in falciparum malaria in Fengqing County and Yun County of Yunnan Province, with the *huanghaosu* extracted by the Yunnan Institute of Materia Medica. Liu Pu from the Beijing Institute of Chinese Materia Medica joined them as an observer. Unfortunately there were very few patients to study. In October that year, the national Project 523 head office director, Zhang Jianfang, researcher Shi Linrong, the Guangdong 523 office director, Cai Hengzheng, and the Kunming 523 office deputy director, Wei Zuomin, went to Yunnan to inspect the progress of the fieldwork. At the clinical study site in Yun County, the four inspectors realized that it was difficult to get enough patients to enter into a study. There were at that time only two vivax malaria cases and one falciparum malaria case. Director Zhang Jianfang and the other inspectors then went to Gengma County Hospital to review the cerebral malaria treatment research by Li Guoqiao's team. At that time, falciparum malaria was epidemic in Lima District, with many cerebral malaria patients. Director Zhang Jianfang sought Li Guoqiao's opinion and hoped that Li's team would take on the clinical study with *huanghaosu*. Li Guoqiao accepted the task. Director Zhang Jianfang then requested the Yunnan *huanghaosu* clinical team to visit areas with a high malaria incidence to collect patients into the study. He also requested the Yunnan clinical team to immediately deliver some *huanghaosu* to Li Guoqiao in Lima District to carry out the clinical trial. On returning to Kunming, Director Zhang Jianfang made the necessary arrangements with Director Fu Liangshu of the Kunming Project 523 local office to proceed with the studies.

Following the instructions of the national Project 523 head office, three members of the Yunnan clinical study team (Lu Weidong of the Yunnan Institute of Materia Medica, laboratory technician Wang Xuezhong of the Institute of Malaria Prevention, and an observer from the Beijing Institute of Chinese Materia Medica) delivered *huanghaosu* to Li Guoqiao's group in Lima to start a clinical trial. The effectiveness of the drug observed on the first falciparum patient entered into the study surprised Li Guoqiao. After the second patient, Li Guoqiao told Lu Weidong and others that he had not seen antimalarial drugs with such rapid responses and would like to try the drug on cerebral malaria patients as soon as possible. The third patient also had good results. By mid-November, the Yunnan clinical study team returned to Kunmimg. Li Guoqiao and his team continued their study in cerebral malaria. The clinical trial with *huanghaosu* in falciparum malaria was completed very quickly with very satisfactory results. From this point on, a new page had been turned in the research on *huanghaosu*!

***

E. Good News from Lima on Treating Falciparum Malaria

It was November 1974 and the end of the fieldwork phase when Li Guoqiao's team accepted the task of undertaking clinical trials with *huanghaosu*. Because of the urgency to evaluate *huanghaosu* in falciparum malaria, the team stayed on to continue the clinical work. From the first three cases, Li Guoqiao observed that after drug administration the falciparum parasites at the ring form stage of development could not mature into the large trophozoite forms, and he felt that the action of *huanghaosu* against falciparum malaria parasites was very quick, far exceeding the activity of quinine and chloroquine. Although *huanghaosu* was not water-soluble, and could not be used by the intramuscular or intravenous injection route, Li felt that if the crushed *huanghaosu* tablets were mixed with water, the drug could be administered to comatose cerebral malaria patients via a feeding tube through the nose. There was the possibility that parasites might be destroyed faster in this way than by using intravenous quinine. Li immediately went to test this hypothesis in the field, in the Nan La community of Cangyuen County at the province border.

Li Guoqiao's team continued to work until January 1975. In addition to the original three cases, they had entered fifteen malaria patients from the Gengma County Hospital and the Nan La Community Health Institute of Cangyuen County. The total patient number reached eighteen, with fourteen having falciparum malaria (including three with malignant malaria) and four with vivax malaria. The results were very positive.

The decision to assign the clinical trial of *qinghaosu/ huanghaosu* extracted by the Yunnan Institute of Chinese Materia Medica to Li Guoqiao's team was based on their knowledge and large clinical experience in treating malignant malaria, especially cerebral malaria. For seven consecutive years, Li's team had been treating falciparum malaria in Hainan, Guangdong Province, and in Yunnan Province. Li Guoqiao had followed the cycle of *in vivo* parasite development in patients, observed the post-treatment parasite morphologic change, and correlated these observations with clinical manifestation in the patients. Li was very familiar with the disease falciparum malaria and the characteristics of drug effects on the parasites. He had a unique method of studying the effectiveness of drugs on parasites. Clinical trials of the most important drugs from Project 523 were assigned to Li's team. In the clinical trial comparing *qinghaosu/huanghaosu* versus chloroquine, Li applied his parasite observation method to nine of the eighteen *qinghaosu/huanghaosu* patients (and five chloroquine patients). For both treatment groups, drugs were given when the parasites were at the tiny ring-form stage of parasite development. Four-hourly blood smears were taken for a parasite count and to record morphology changes. Six blood smears were studied within the first twenty-four hours. In the first twenty-four hours in the *qinghaosu/huanghaosu* group, parasite counts started declining six hours after drug administration, with a 90 percent clearance of parasites at sixteen hours and more than 95 percent parasite clearance at twenty hours. In the chloroquine group, parasite counts started declining at twelve hours with an average of only 47 percent parasite clearance at sixteen hours and a 95 percent clearance occurring at only forty hours. This showed that *qinghaosu/huanghaosu* cleared falciparum malaria parasites faster than chloroquine. But Li also found that there was a problem with reappearance of parasitemia and recurrence of symptoms after the treatment with *qinghaosu/huanghaosu*. The following is an excerpt from Li's report, "Treating 18 malaria cases with *qinghaosu/huanghaosu*: A summary."

"From October to December 1974, 18 malaria cases were treated with *qinghaosu/huanghaosu* in the Gengma County Hospital and in the Nan La Community Health Institute of Cangyuen County. Fourteen cases were falciparum malaria (including three with malignant malaria) and four had vivax malaria. Preliminary results showed that *qinghaosu/huanghaosu* was faster than chloroquine in eliminating malaria parasites.

***

Treatment effects

1. Temperature normalization time.

In 14 falciparum cases, one case did not have a high temperature, while the average temperature normalization time in 13 feverish patients was 37.5 hours.

2. The speed of parasite destruction.

In the *qinghaosu/huanghaosu* group detailed blood smear analysis was performed in nine falciparum malaria cases to study the rapidity of parasite destruction. Drugs were given at the tiny-ring form stage of parasite development, and it was seen that parasite counts started declining six hours later, with a 90 percent parasite clearance at sixteen hours, and greater than a 95 percent parasite clearance at twenty hours. Also parasite development was arrested at the tiny-ring form stage following drug administration.

In five falciparum malaria cases in the chloroquine phosphate treated group, two received it intravenously and three orally. All patients received 30mg/kg on day 1 (oral group received the total dose within 6 hours; IV group within 8 hours.), 10mg/kg on day 2 and 10mg/kg on day 3. Drugs were given at the ring form stage of parasite development. Parasites started declining at 12 hours after drug administration with an average clearance at 16 hours of 47%, a clearance at 24 hours of 74%, and 95% parasite clearance at 40 hours. Much later after drug administration, the remaining parasites continued to mature from the tiny-ring forms into medium-sized ring forms or the thick, large-ring forms.

These results demonstrate that *qinghaosu/huanghaosu* clears malaria parasites faster than chloroquine.

One vivax case was studied with this detailed blood smear analysis method. *Qinghaosu/huanghaosu* 0.6 grams was given when the parasite nucleus was divided into 2 to 3 nuclei. In the period from 2–7 hours after drug administration, hourly blood smears were made. Most parasites had maturation arrest at the stage with 2–3 nuclei with rare maturation to the stage with 4–5 nuclei. This showed that *qinghaosu/huanghaosu* affected vivax malaria parasite development two hours after drug administration.

3. Time to achieve parasite clearance.

In the 14 falciparum malaria cases, two that failed treatment were not followed with detailed malaria smear examinations. In the 12 cases that were studied, the shortest time to achieve a negative parasite count was 28 hours, and the longest 84 hours, with an average time of 54 hours. One of the two failed cases was treated with *qinghaosu/huanghaosu* 1.5 grams for 1 day only. Parasitemia persisted and was present in the blood smear at 68 hours, and symptoms recurred 8 days after drug administration. The other failed case was a patient with malaria and jaundice and an initial parasite count of 334,400/mm^3. A 2-day treatment regimen with *qinghaosu/huanghaosu* (1 gram on day 1, 0.5 grams on day 2) did not clear the parasitemia which was still present on day seven after drug administration.

Negative parasitemia was achieved in all four vivax patients with the shortest time being 40 hours, and the longest time 48 hours, with an average of 43.5 hours.

4. The reappearance of parasitemia and recurrence of symptoms.

Seven falciparum malaria cases had a short-term follow-up. Six had positive blood smears and symptoms recurrence on day 8 to day 24 after drug administration; one case had a negative blood smear on day 11 but was lost to follow-up.

5. Details of one case of malignant falciparum malaria treated with qinghaosu/huanghaosu.

A female 20 years old of the Dai minority group was 6 months pregnant with intra-utero fetal death on day 10 of the disease. She had severe anemia (red blood cells 1.2 million /mm^3), a body temperature below 35°C with no pre-admission antimalarial drugs given. A blood smear showed thick large-ring forms of falciparum parasites with a count of 223,440/mm^3 and maturing into large trophozoites in the capillaries. The patient was non-comatose. She delivered a dead fetus after admission to the obstetrics department. The patient had all the criteria of severe malignant falciparum malaria: severe anemia, low body temperature, spontaneous abortion, large number of parasites collected in capillaries in the internal organs. Because the patient was not in coma, 0.5 grams of *qinghaosu/huanghaosu* was given orally together with supportive treatment and a blood transfusion. When 0.5 grams of *qinghaosu/ huanghaosu* was given 4 hours later, the patient was confused with blurring of speech and she became comatose shortly after. At this time more than 90% of the thick large ring forms had left the peripheral blood and collected in organ capillaries. On day two the patient was still in coma. From observations on some 10 cases previously treated with *qinghaosu/huanghaosu*, and considering that this patient had already had 1 gram of *qinghaosu/huanghaosu* orally, it was felt with confidence that the parasites would be controlled and no more antimalaria drugs were needed. (Also in part because the patient was in coma and could not take any more oral drug). Supportive treatment for cerebral edema and other complications was given as well as further blood transfusions. The patient came out of coma 50 hours later, and a blood smear was negative at 72 hours.

6. Side effects:

Vomiting occurred in the two cases receiving 2 grams on day 1. No clinical side effects were observed in all the other patients.

In nine cases (seven falciparum, two vivax), eight had normal liver function tests on day 3–6 after drug administration; one case did not have baseline liver function tests, and on day 3 after the drug the liver enzyme (ALT) was high at 200 units/ml, which persisted at 190–200 units/ml from day 3–6. Whether this was due to the large drug dosage is unknown.

7. Conclusions:

This preliminary clinical trial demonstrates that: a) *Qinghaosu/ huanghaosu* rapidly destroys malaria parasites giving a short parasite clearance time, but parasites reappear and symptoms recur quickly; b) The speed of parasite clearance was not dose dependent in the range of 0.25 to 1 gram for the first dose, and 0.5 to 2 grams for the first day total dose; c) Parasite counts were lowered rapidly in vivax malaria patients taking 0.2 to 0.3 gram as a single dose, and negative blood smears were achieved within 40–48 hours. Preliminary impression: *qinghaosu/ huanghaosu* is a fast acting antimalarial drug with rapid control of malaria parasite development which could be achieved with a first dose of 0.3 to 0.5 gram.

The relatively quick reappearance of parasites and the recurrence of symptoms may result from the short duration of an effective blood level due to the rapid elimination of drug from the body, or to rapid metabolism to an inactive substance, so that not all the initial parasites are destroyed. This problem may be overcome by altering the dose and duration and timing of administration of the drug, in future trials.

Although *qinghaosu/huanghaosu* had a questionable long lasting effect, it was certainly fast acting and effective in treating malignant malaria. It is suggested that the production of a *qinghaosu/huanghaosu* injection form be developed as soon as possible for clinical use."[24]

***

Observing the rapid parasite-destroying characteristic of *qinghaosu/huanghaosu*, Li Guoqiao took the initiative and decided that such an action could be important in treating cerebral malaria, a frequently fatal disease. Li's team fed crushed tablets through a nasogastric tube and successfully treated three patients with cerebral malaria, one of whom was pregnant. This was a bold experiment with enormous significance for the continuing development of *Qinghao* and its potential future use.

***

With the *qinghaosu/huanghaosu* provided by the Yunnan Institute of Materia Medica and the clinical trials in falciparum malaria carried out in Gengma County by the Guangzhou College of Traditional Chinese Medicine (now Guangzhou University of Chinese Medicine), the basic characteristics of *qinghaosu/huanghaosu* were demonstrated: rapid onset of action, high short-term cure rate, low toxicity, no early drug resistance, but with a high recrudescence rate. These findings led to the decision to evaluate the effectiveness and toxicity of *qinghaosu/huanghaosu* in treating drug-resistant falciparum malaria in tropical areas. The Shandong Institute of Chinese Materia Medica, collaborating with the Institute of Parasitic Diseases, also confirmed the effectiveness, safety, but high recurrence rate of *qinghaosu/huanghaosu* in a clinical trial with nineteen vivax malaria cases. All these findings were confirmed later in many studies in and outside China. The rapid effect, high cure rate, and low toxicity characteristics of *qinghaosu/huanghaosu* are hard to surpass or exceed by any other antimalarial drugs to date.

The confirmation of its effectiveness against drug-resistant falciparum malaria served especially as the cornerstone for the 1975 decision of the national Project 523 leading group, to concentrate all efforts from all fronts nationwide in the further development of *qinghaosu/huanghaosu*.[25]

CHAPTER 3

Collaboration between Organizations Involved in Accelerating the Development Program

In 1975, strategic decisions were made to concentrate on *Qinghao* and *qinghaosu* and to stop research on other compounds. Goals were set through 1977 to expedite the development of antimalarial drugs based only, therefore, on *Qinghao* and *qinghaosu*.[25] Several meetings at which progress was reviewed were central to the decision-making and planning processes.

***

A. The Beijing Beiwei Road Hotel Meeting

At the end of February 1975, the national Project 523 head office called a meeting of the leaders of all Project 523 district offices and task units at the Beiwei Road Hotel in Beijing. Progress on *qinghaosu* extraction by various units and clinical trials was presented.[26, 27] Impressive results from eighteen patients treated with *qinghaosu* by Li Guoqiao in Yunnan the previous month and supporting data from a number of sites over the previous three years led to the decision to direct more research units to work on *qinghaosu*, as mentioned earlier, and to make further development of the drug the main target for the 1975 research plan.[25]

***

By 1974 the Beijing Institute of Traditional Chinese Materia Medica was still unsuccessful in extracting *qinghaosu*.[26, 27] In 1975, with arrangements by the national Project 523 head office researchers from the foregoing institute were sent to Sichuan Institute of Traditional Chinese Materia Medica to learn the technique for extraction. With improved techniques, Beijing Institute was able to extract 500 grams of *qinghaosu* by May 1975, enough supply for subsequent clinical, chemical structural, and pharmacological studies at the Beijing Institute of Traditional Chinese Materia Medica.[28]

***

B. The Chengdu Meeting

In April 1975, a meeting was held in Chengdu, attended by seventy team leaders and researchers from the various national ministries and commissions, Military Headquarters, and members of thirty-nine research teams from ten provincial, city, district, and military affiliated organizations. Participants reported on their work. The most encouraging data were from the studies of Li Guoqiao in Yunnan, on uncomplicated and malignant falciparum malaria, and from the studies of the Shandong *Huanghuahao* research collaborative group on *P. vivax* malaria.

To expedite research on Q*inghao* nationwide, the meeting arranged cooperation between specialty groups and assigned research projects to appropriate individual research teams.

The main tasks in *Qinghao* research identified by the Chengdu meeting were as follows:

1. Locate good sources of *Qinghao* plants, and use locally available materials to manufacture a simple

preparation of *Qinghao* that was easy to use, inexpensive, and effective;

2. Improve the technique for extracting *qinghaosu* from *Qinghao*, increase the effectiveness of the resulting drugs, and decrease the recurrence rate;

3. Identify the chemical structure of *qinghaosu* and possibly modify it, continue pharmacological, toxicological, and drug metabolism studies, and develop different pharmaceutical forms for administration (e.g., oral, intramuscular, and intravenous).[29]

This Chengdu meeting brought medical, pharmaceutical, and other specialty research groups in China to work together, and was the driving force behind future *Qinghao* research. It was also the first truly collaborative meeting, uniting all units to work towards a common goal.

***

C. Establishing the Chemical Structure of Qinghaosu

By the time of the 1975 meetings, the chemical structure of *qinghaosu* was still not known, despite considerable effort. In 1972, after animal studies and clinical trials with crude ether extracts of *Qinghao* had shown promising results at the Beijing Institute of Chinese Materia Medica, the Project 523 head office arranged for the isolation of the active monomers of *Qinghao* and the identification of its chemical structure.[11] The Beijing, Shandong, and Yunnan Institutes of Chinese Materia Medica expedited their research in extracting the active crystals and did preparatory work towards establishing the chemical structure. The aim was to identify the structure of the active *Qinghao* monomer and then synthesize it by 1974.[30]

Establishing the chemical structure of *qinghaosu* was a collaborative effort between the Beijing Institute of Chinese Materia Medica and the Shanghai Institute of Organic Chemistry of the Chinese Academy of Sciences. The Shanghai group included Wu Zhaohua and Wu Yulin, under the supervision of Professor Zhou Weishan.[31] The Beijing group included Liu Jienmien, Fan Jufen, and Nie Muyun. Because the Beijing Institute was unable to supply *qinghaosu* extract for the chemical structure study as planned, the national Project 523 head office asked the Shandong and Yunnan Institutes to provide the Beijing Institute with their purer *qinghaosu* crystals for the work on the chemical structure.

Initial progress was slow. In early 1975, Professor Zhou Weishan from the Shanghai Institute visited the Yunnan Institute. While there, he acknowledged that the identification was difficult and the supply of *qinghaosu* was insufficient for the research work. He said that the researchers from the Beijing Institute had returned to Beijing, uncertain whether they could continue with the project.[32]

The Shanghai Institute was using a method of chemical structure identification that involved elemental analysis, deoxidization, hydrogenation, and then chemical structure identification of the substance obtained by circular dichromism and mass spectrometry. Meanwhile, the Beijing Institute also conducted chemical reaction studies under the guidance of Professor Liang Shaotien. With the help of other scientific research units, using a 250 MH2 nuclear magnetic resonance (NMR) machine, the hydrogen and carbon spectra of *qinghaosu* were obtained, which helped to establish the parent nucleus structure. Extrapolating from the fact that the *yingzhaosu* chemical structure contained a peroxide group, the chemical molecular structure of *qinghaosu* was identified, and the relative configuration of the chemical structure was then established. With the confirmation that *qinghaosu* was a kind of peroxide, a series of chemical reactions (such as hydrogenation-deoxidization) of *qinghaosu* was performed. These showed that the lactone of *qinghaosu* could be reduced into a pair of isomers of dihydro-artemisinin by sodium or potassium borohydride and with the preservation of the peroxide group. [33] Further experimentation with various chemical reactions was the foundation for the future development of *qinghaosu* derivatives.

Later, researchers Liang Li and others from the Institute of Biophysics, Chinese Academy of Sciences, used the method of x-ray single crystal diffraction to establish the absolute configuration of *qinghaosu*. It was now established that *qinghaosu* was formed from carbon, hydrogen, and oxygen, and was a new sesquiterpene lactone containing a peroxide group, which was a totally new chemical compound with a structure totally different from all the known antimalarial drugs.

***

D. Pursuing the Leads and Expanding the Research

Even as efforts to identify the chemical structure of artemisinin were ongoing, clinical and industrial work continued with the aim of completing Project 523 as quickly as possible. *Qinghaosu* clinical trials directed by the Beijing Institute of Traditional Chinese Materia Medica were carried out in Hainan, Guangdong Province, and in Hubei and Henan Provinces. The *qinghaosu* used in these clinical trials was manufactured by Wuhan Jianming Pharmaceutical Factory in Hubei Province from *Qinghao* plants from Youyang District of Sichuan Province. A solvent-gas extraction method, developed in Yunnan Province, was used. Meanwhile, the Beijing Institute of Materia Medica also helped Gaoyou County of Jiangsu Province undertake clinical trials with the crude preparation of *Qinghao* plants, further confirming its efficacy.[15]

Cooperation between multiple research units also continued on the pharmacology, toxicology, chemistry, and pharmaceutical aspects of *qinghaosu* development.

***

The Second Beijing Meeting of 1975

To follow up after the Chengdu meeting and to make additional plans, the national Project 523 head office called another meeting in Beijing at the end of 1975. Local reports further confirmed the rapid action of *qinghaosu*, its high short-term cure rate and low side effect rate. Still remaining were the high recurrence rate, high extraction cost, and still unclear pharmacology. The following priority projects for 1976 were decided upon:

1. Improve and validate simple preparations of crude, unpurified *Qinghao* and then choose a few preparations with the best cure rate, lowest side effects, and the least expensive to obtain, and promote these for wider use;
2. Design treatment protocols for uncomplicated and malignant malaria and train clinical units to test and validate the protocols; perform larger clinical trials on malignant malaria;
3. Develop preparations suitable for parenteral injection (intramuscular and intravenous);
4. Improve the plant extraction technique and manufacturing process to obtain a high yield as efficiently as possible and at a low cost;
5. Determine the exact chemical structure and correlate this with its mode of action in order to investigate and develop newer drug forms.

After discussions at the meeting and with the agreement of the national Project 523 head office, the Shanghai Institute of

Materia Medica was assigned the task of modifying the chemical structure of *qinghaosu* to make new derivatives.[34]

At the meeting in late 1975, Project 523 head office decided it was time to develop a range of drug forms. It therefore invited the two institutes capable of extracting large quantities of *qinghaosu* (Shandong Institute of Chinese Traditional Medicine and Yunnan Institute of Materia Medica) to prepare the drug for administration as tablets and capsules, and as aqueous and oil emulsion injection forms.

At about the same time, Cambodia asked the Chinese government to send a "Malaria Prevention and Treatment Observation Group" to help control malaria. From January through July 1976, a team went to Cambodia with *qinghaosu*, the first time the drug was used outside China. *Qinghaosu* was shown to be effective against drug-resistant falciparum malaria as well as cerebral malaria.[35]

***

E. Gathering Data for Validating the Results of Qinghaosu

In July 1976, the national Project 523 head office called a meeting in Gaoyou County of Jiangsu Province. The field experience of the Department of Health in Gaoyou County led to further adjustment of the *Qinghao* research plan. It was decided that larger clinical trials with simple preparations of crude, unpurified *Qinghao* were called for to confirm and improve efficacy. The most effective one or two preparations would then be used on an even larger scale. Criteria for selection of the best would be based on local availability of the raw plant source and facilities to manufacture the drug using techniques appropriate to local villages. By 1977 expanded crossover trials between districts must be organized.

The head office also wanted to prepare for the drug evaluation and registration process. In addition to confirming the chemical structure, pharmacology, toxicology, and drug metabolism studies had to be completed. Meanwhile research on structure modification for improved drugs would continue, as would increased efforts to improve production, extraction, purification, and use of the then-available preparations.[36]

In December 1976, the Project 523 head office asked the Beijing Institute of Traditional Chinese Materia Medica and the Shandong Institute of Chinese Traditional Medicine to organize a training and information exchange course. One goal was to decide on laboratory methodology and standards for *qinghaosu* quality determinations. Different preparations had been available and in use throughout China for many years, but it was now time to consolidate the data and to standardize the product to be evaluated and registered.

The two-week training course was held in February 1977 in Shandong. Participants included the Beijing Institute of Materia Medica, Shandong Institute of Chinese Traditional Medicine, Yunnan Institute of Materia Medica, and fifteen others from various provincial or city institutes of material medica, botany, drug evaluation, parasitic diseases, and pharmaceutical factories, from Shanghai, Guangdong, Guangxi, Jiangxu, Henan, Sichuan, and Hebei. Participants from all the relevant institutions brought 1–2 kg of *Qinghao* plants from their own locality to be tested.[37] Participants from each institute presented their own techniques of testing *qinghaosu* concentration (about ten methods), and then used these techniques to test the *Qinghao* plants and *qinghaosu* provided by Sichuan and Shandong Provinces. The pros and cons of each technique and its practicality were analyzed and discussed. The ultraviolet spectrophotometric methods developed by Nanjing Pharmaceutical College and Guangzhou College of Traditional Chinese Medicine were deemed the best methods, but it was felt they could be improved. The training and exchange course then assigned further research tasks to various teams to establish a standard quality reference for *qinghaosu* and *Qinghao* preparations that could be used to test the concentration of *qinghaosu* in *Qinghao* and *qinghaosu* content in their pharmaceutical preparations.[38]

In September 1977, the Project 523 head office requested a second *qinghaosu* concentration testing training-exchange course at the Beijing Institute of Chinese Materia Medica. The agenda of this meeting was: 1) To clearly define a reliable and practical *qinghaosu* concentration testing method; 2) To examine the *qinghaosu* quality reference established independently by the Beijing Institute of Traditional Chinese Materia Medica, Shandong Institute of Chinese Traditional Medicine, and Yunnan Institute of Materia Medica, and to decide on a single *qinghaosu* standard quality reference. Because of the requirements for entry in the China Pharmacopoeia, a representative of the Central Institute for Drug Control was also invited to the meeting. Regarding the concentration testing method, participants still felt that the ultraviolet spectrophotometric method developed by Nanjing Pharmacology College and Guangzhou College of Traditional Chinese Medicine was the best, but the improvement suggested by the last training-exchange course was still not completed. Possible solutions suggested at the meeting were tested immediately at the Beijing Institute of Traditional Chinese Materia Medica, and a final acceptable concentration testing method was established. A draft of the method was prepared by the Beijing Institute of Chinese Materia Medica, supplemented by the data from the Shandong Institute of Chinese Traditional Medicine and the Yunnan Institute of Materia Medica. The final document was written by five participants (Zeng Meiyi, Tian Ying, Luo Zeyuan, Shen Xuankun, and Yen Kedong).

***

Final Steps

In May 1977, the Project 523 head office called a meeting in Nanning, Guangxi Province, to thoroughly prepare for the formal evaluation of *qinghaosu* and *Qinghao* simple preparations. The "Combined Western and Traditional Medicine Specialty Group Meeting on Antimalarial Drugs" (the "Nanning Meeting") summarized and evaluated the results and progress on *Qinghao* and *qinghaosu* research during the two years since the 1975 Chengdu meeting. The meeting emphasized tasks that needed to be completed before the formal evaluation process could be initiated.

Since the 1975 Chengdu meeting and after three major national meetings during the two-year period, clinical trials in Jiangsu, Sichuan, and Guangdong had used *qinghaosu* and *Qinghao* simple preparations in treating two thousand malaria patients. *Qinghao* simple preparations were used in 1,200 cases, with parasite clearance rate above 90 percent. Various preparations of *qinghaosu* were used in eight hundred cases, with a parasite clearance rate of 100 percent. But with both *Qinghao* preparations and *qinghaosu*, the one-month recrudescence rate was about 30 percent, clearly still not satisfactory. Yunnan and Shandong used a *qinghaosu* injectable form, and the one-month recrudescence rate decreased from 30 percent to below 10 percent, but the problem of a high recrudescence rate was still not satisfactorily solved. In eighty-four cerebral malaria cases, *qinghaosu* was more effective than quinine or other available antimalarials.

The chemical structure of *qinghaosu* was established, and a few derivatives were made. The research on *qinghaosu* pharmacology, toxicology, and metabolism was started. *Qinghaosu* concentration testing with ultraviolet spectrophotometric technique was decided upon. The *qinghaosu* standard quality reference was established. Improvement was achieved on *qinghaosu* extraction techniques and production processes, and different pharmaceutical formulations had been developed.

The Nanning meeting ended with a resolution to complete all data and supplementary information requirements in order to start the drug evaluation and registration process in 1977–1978. Production of urgently needed antimalarial drugs should begin as soon as possible.[39]

CHAPTER 4

A Historical Review

Major progress was made in the research and development of *Qinghao* and *qinghaosu* over three years, from 1972 to 1975. In 1972, the Beijing Institute of Traditional Chinese Materia Medica reported successful treatment of mice with *Plasmodium berghei* using ether extracts of *Qinghao*. By 1975, Professor Li and colleagues reported the successful use of *qinghaosu* for treating drug-resistant falciparum malaria in patients in the highly endemic area of Yunnan Province. In another three years, after additional multidisciplinary and multi-unit cooperative studies, the drug completed the formal evaluation process in 1978. During these six years, a great deal of difficult and often tedious work was completed, and success was achieved as a result of the collaborative efforts of many research departments throughout the country, under the direction of a central coordinating Project 523 national head office in Beijing. Such rapid progress would not have been possible with a single research unit, however capable. The resulting drug, *qinghaosu*, was a new type of antimalarial that would in time achieve an international reputation. In March 1977, the Ministry of Health, Ministry of Chemical Industries, General Logistics Department of the People's Liberation Army, and the Chinese Academy of Sciences held a meeting to address all local Project 523 leaders of provinces, cities, districts, departments, commissions, or military affiliated units. The meeting report said:

"Research on the antimalarial effect of *Qinghao* in only two short years has brought excellent results. The drug extract has been tried on over 3,000 patients and its effectiveness has been confirmed. The research proceeded further to involve *Qinghao* resource investigation, extraction from plants by chemical methods, pharmacology and toxicology, elucidation of the chemical structure, the manufacturing process, measurement and standardization of active content, and pharmaceutical formulations. A single department or a single research unit alone could never have achieved this rapid progress."[40]

A report prepared for a subsequent meeting, the *Qinghao* Research Evaluation Meeting in October 1977, commented as follows:

"After 10 years of hard work *qinghaosu* research has reached a level acceptable to national and international authorities as a standard for treating cerebral malaria and chloroquine-resistant falciparum malaria."

The report was in four parts. Part 1 described the development and research process of *Qinghao* and *qinghaosu*. It begins in 1972 with the Beijing Institute of Traditional Chinese Materia Medica using a simple *Qinghao* extraction in a clinical trial, with good results. But in a September 1973 clinical trial with *Qinghao* crystals, the product was ineffective and had side effects. This setback slowed progress of the research.

Researchers from the Shandong Institute of Parasitic Diseases, having read about the *Qinghao* research in the Nanjing Chinese medicinal herbs specialty group meeting report, used *Huanghuahao* extracts from their own locality (Shandong Province) to treat thirty vivax malaria patients and had good results. The institute then collaborated with the Shandong Institute of Chinese Traditional Medicine to isolate the effective monomer and named it "*huanghuahaosu.*" In 1974, *huanghuahaosu* was tried on twenty-six vivax cases with even better effects.

Independently in Yunnan, in April 1973, the Yunnan Institute of Materia Medica extracted *huanghaosu* directly using chemical solvents from local *Huanghuahao* with large flower heads. Pharmacology and toxicology studies reported low toxicity of this extract. From November 1974 to January 1975, the clinical team from Guangzhou College of Traditional Chinese Medicine used this extract in eighteen patients with malaria (some with falciparum, some with vivax) and found it had a faster onset of action and higher cure rate than chloroquine. They commented that the combined quality of rapid action and high cure rate of *huanghaosu* exceeded any currently used antimalarial drugs. The national Project 523 head office felt that "the clinical trials of forty-four malaria cases in Yunnan and Shandong have supported the decision to continue further research on *Qinghao*."[16]

Parts 2 and 3 of the *Qinghao* Research Evaluation Meeting report described *Qinghao* research progress and cited some remaining problems to be solved. Part 4 made suggestions for the next steps in the research program.

The next major event was the *Qinghao* Research Results Evaluation Conference in March 1978. To prepare for this meeting, the national Project 523 head office organized a pre-evaluation preparation meeting. They gathered research data from all research units and teams, grouped them into twelve topics, and assigned each topic to a group to examine the relevant data and draft a report to be presented at the meeting. During drafting of the "Report for Evaluation" at this meeting, one of the issues was the listing order of involved units, a problem resolved after further discussions. (Chapter 5 has a detailed discussion of this report and meeting.)

***

A. *Pharmacology-Pharmaceutical Studies*

The development of various *qinghaosu* formulations was geared towards ease of use and an attempt to decrease the recrudescence rate, which was a consistent problem. Five oral preparations were developed: powder, tablet, solid dispersant, microcapsule, and pills. The three parenteral forms, intended to address the high recrudescence rate, were oil injection, oil emulsion injection, and aqueous injection.

The simple unpurified crude *Qinghao* preparations studied were fresh squeezed juice, decoction, granules, and compressed extraction in tablet form. Some teams studied combination regimens of *Qinghao* simple preparation with other drugs and suggested regimens feasible for mass treatment and prevention in the general population in endemic areas.

The best methods for determining the *qinghaosu* content and quality of the preparations were the subject of two training-information exchange courses. It was determined that ultraviolet spectrophotometry was the best method for *qinghaosu* content determination, and the quality standard for *qinghaosu* was based on data provided by the Beijing Institute of Traditional Chinese Materia Medica, complemented by information from Shandong and Yunnan.

***

B. Clinical Trials

The efficacy of *huanghaosu* (*qinghaosu*) had been shown in thirty-seven malaria cases in Shandong and Yunnan in 1974, but confirming usefulness, especially for uncomplicated falciparum malaria and severe malaria, was limited because the case numbers were small and limited to specific areas. The Chengdu meeting suggested expanded clinical trials in each locality, which each unit undertook using various *huanghaosu* preparation forms produced locally with local *Huanghuahao* (*Huanghao*) plants. The cure rate of *huanghaosu* was 100 percent when tested in Yunnan, Guangdong's Hainan Island, and Cambodia. Sichuan, Guangdong, and Henan also had 100 percent cure rates with the *Qinghao* tablets they formulated. In these expanded trials, a total of 6,550 malaria cases of various forms were treated effectively with *qinghaosu* (2,099 cases) and simple *Qinghao* preparations (4,451) in Guangdong (Hainan), Yunnan, Henan, Shandong, Jiangsu, Hubei, Sichuan, Guangxi, and in Laos and Cambodia.[4] *Qinghaosu* in oral form or in injection form had a short term cure rate of 100 percent; *qinghaosu* oil emulsion injection form had a recrudescent rate in one month of about 10 percent. The cure rate of *Qinghao* simple preparation was over 80 percent.

***

C. *Qinghaosu Resource Distribution Mapped*

After the Chengdu meeting, scientists with expertise in botany and raw preparations of Chinese herbs from Guangdong, Yunnan, Jiangsu, Sichuan, Guangxi, Fujian, and Beijing quickly investigated the distribution of wild *Qinghao* in China. They found that *Qinghao* grew in abundance in large areas in many northern and southern provinces. Only a small proportion was sold as medicinal herbs while the majority of the material had been used as fertilizer. The Guilin District of Guangxi Province could produce five hundred tons per year, and Guangdong five thousand tons per year. In general the *qinghaosu* content in the raw *Qinghao* herb went from higher to lower moving from south to north. The high *Qinghao* content areas were located south of the Nanling and the Wuyi mountain ranges, with the highest content in Guangxi and Guangdong Provinces, especially in the northern part of Hainan Island. Moderate *Qinghao* content was found between the Nanling Range in the south and the Qinglin Range in the north. In this middle region (especially in the Youyan, Pengsui, and Siushang Districts of Chungqing, the Shangsi District and Wulin mountainous district of Hunan, and at the borders between Guizhou, Yunnan, Guangxi, and Hunan),

Huanghuahao grew in abundance and had a reasonable level of *Qinghao* content. The low *Qinghao* content area lay north of the Qinglin Range, and *Qinghao* content was lowest in the northeast region of China.

***

D. *Pharmacological Characteristics*

Pharmacology and toxicology information on *qinghaosu* came from laboratory experiments and clinical trials carried out by the Beijing Institute of Traditional Chinese Materia Medica, the Shandong Institute of Chinese Traditional Medicine, the Yunnan Institute of Materia Medica, and the research units from Guangdong, Guangxi, Sichuan, and Jiangsu Provinces. The rapid clearance of malaria parasites with *qinghaosu* was faster than with chloroquine and all other antimalarial drugs available at the time. But the recrudescence rate was also high, consistent with *qinghaosu*'s rapid absorption, wide distribution, rapid excretion, and consequent lack of persistence in the body.

Qinghaosu was also determined to be a drug of low toxicity. In both laboratory experiments and in clinical trials, *qinghaosu* in the therapeutic dose range had no apparent adverse effects on the heart, liver, or kidneys. No side effects were observed when given to pregnant patients. No abnormalities were seen in animals either. *Qinghaosu* was effective in all patients in whom chloroquine had failed in Yunnan and Guangdong (Hainan Island). This finding signified that *qinghaosu* had no cross–drug resistance with chloroquine and could therefore be used in treating chloroquine-resistant malaria. Because *qinghaosu* clears parasitemia very quickly, it is the drug of choice for treating cerebral malaria and other forms of severe malaria.

***

E. Extraction Techniques

After the Yunnan Institute of Materia Medica improved *qinghaosu* extraction by a solvent-gas method in 1974, it collaborated with Kunming Pharmaceutical Factory in more expansive experiments to manufacture *huanghaosu*. After laboratory experimentation, two large-scale manufacturing experiments with solvent-gas extracted *qinghaosu* were carried out, the first from December 1975 to January 1976, and the second from September to October 1976. In these expanded experimental manufacturing processes, 2.256 tons of *huanghuahao* dried leaves were used for extraction of *huanghaosu* with benzine 120. This manufacturing process had all the desired characteristics: short flow times, high yield, low capital cost, easy purification, simple operation, steady production process, no need for special equipment or testing agents, and with end products of high purity. This eventually became the standard production method in other places.

The Yunnan Institute of Materia Medica's extraction process for batch production of *huanghaosu* contributed significantly in providing sufficient material for research and field studies, and for use by the military. The acetone extraction technique later developed by the Shandong Institute of Chinese Traditional Medicine increased the yield of *qinghaosu* from low-concentration *Qinghao*. Guilin Aroma Chemical Factory in Guangxi modified the Yunnan solvent-gas extraction technique and increased the yield of *qinghaosu* in Guilin District.

***

F. Chemical Structure

The chemical structure of *qinghaosu* was elucidated by the Shanghai Institute of Organic Chemistry (with major contributions by Professor Zhou Weishan and researchers Wu Zhaohua and Wu Yulin) and the Chinese Academy of Sciences, with the collaboration of the Beijing Institute of Traditional Chinese Materia Medica, and the Institute of Biophysics of the Chinese Academy of Sciences. Repeated chemical experiments and spectrometry studies were required to establish the molecular configuration of *qinghaosu*. The Institute of Biophysics of the Chinese Academy of Sciences later used an x-ray crystal diffraction method to establish the absolute three-dimensional configuration of the *qinghaosu* molecule. The researchers concluded that *qinghaosu* was composed entirely of carbon, hydrogen, and oxygen and was a sesquiterpene lactone with a peroxide group. *Qinghaosu* was a new chemical compound structurally different from all the known antimalarial drugs.

In 1820 quinine alkaloids were isolated from the bark of the cinchona tree, which had been used by native Peruvian Indians to treat fever. After the chemical structure of quinine was identified in 1908, pharmacologists experimented with compounds with similar chemical structures. Subsequently a series of quinine derivatives was created. Among those with antimalarial activities, exemplified by the 4-aminoquinolines (such as chloroquine), all had a nitrous hetero-ring in their chemical structure. From this observation, scientists concluded that a hetero-ring containing the nitrogen was necessary for an effective antimalarial. The discovery of *qinghaosu* and confirmation of its chemical structure was the first time this concept was proven wrong. *Qinghaosu* was as important a breakthrough as quinine and the quinolones. The discovery also demonstrated the value of exploring the Chinese nation's traditional medicine heritage.

In 1976 information reached China that a phytochemist in Yugoslavia was isolating a substance similar to *qinghaosu* from other species of the *Artemisia* genus. The National Academy of Chinese Traditional Medicine recommended to the Ministry of Health that Chinese data should be published as soon as possible, before publication from Yugoslavia. *Qinghaosu*'s chemical structure was published in the *Chinese Science Bulletin* 1977 under a collective authorship, the "*Qinghaosu*

Structure Collaborative Research Group." In May 1978, in the same journal, the three-dimensional configuration of *qinghaosu* crystals was published under the authorship "*Qinghaosu* Collaborative Research Group," and the Institute of Biophysics of the Chinese Academy of Sciences. In 1979, a second *qinghaosu* chemical structure article was published in *Acta Chimica Sinica* by the researchers from the Beijing Institute of Traditional Chinese Materia Medica and the Shanghai Institute of Organic Chemistry.

CHAPTER 5

Evaluation Committee Approves Qinghaosu (Artemisinin)

After one year of data collection and analysis, the national Project 523 leading group arranged an evaluation meeting on *qinghaosu* for November 23–29, 1978, in the city of Yangzhou in Jiangsu Province. Attending the meeting were leaders and key personnel from the Ministry of Health, the National Commission of Science and Technology, and the General Logistics Department of the People's Liberation Army. Also in attendance were leaders of the participating provinces, cities, districts, and military affiliated districts, as well as the leaders, researchers, and scientists from all the research units involved in the development of *qinghaosu*. Representatives were invited from the Chinese Medical Association, the Pharmacopoeia Committee of the Ministry of Health, the National Institute for the Control of Pharmaceutical and Biological Products, and from the new *Chinese Journal of Medicine*. A total of a hundred and four people attended the evaluation meeting.

***

A. Individual Reports

The *qinghaosu* research project had involved forty-five main research units from various scientific institutions that were directly under the state ministries and commissions, and nine provincial, city, district, or military affiliated institutions such as medical and pharmaceutical colleges and research institutes. Prior to the meeting, twelve detailed reports, covering the relevant topics, had been compiled by fourteen experts. These experts collated the data not only from studies they and their close colleagues had carried out, but also from research units around the country, representing the unstinting efforts of all the researchers involved, most of whom went unnamed. The report topics and authors were as follows:

1. *Qinghao* types and resources, by Wang Guiqing, Guangxi Institute of Botany;
2. Chemistry of *qinghaosu*, by Tu Youyou, Beijing Institute of Traditional Chinese Materia Medica;
3. Pharmacology of *qinghaosu*, by Li Zelin, Beijing Institute of Traditional Chinese Materia Medica;
4. Clinical trials on *Qinghao* preparations in treating falciparum and vivax malaria, by Wang Tongyin, Kunming Medical College;
5. Clinical trials on *qinghaosu* preparations in treating cerebral malaria, by Li Guoqiao, Guangzhou College of Traditional Chinese Medicine;
6. *Qinghaosu* preparations in treating chloroquine-resistant falciparum malaria, by Cai Xianzheng, Hainan District Health and Quarantine Station;
7. Testing *qinghaosu* content and establishing quality standards, by Zeng Meiyi, Beijing Institute of Traditional Chinese Materia Medica;
8. Research on *qinghaosu* formulations, by Tian Ying, Shandong Institute of Chinese Traditional Medicine;
9. Research on the industrial production or process of *qinghaosu*, by Zhan Eryi, Yunnan Institute of Materia Medica;

10. Research on the industrial production process of *qinghaosu*, by Deng Zheheng, Guilin Aroma Chemical Factory;
11. Production process, pharmacology, and clinical study of *Qinghao* extract tablets, by Wu Huizhang and Cang Qizhong, Sichuan Institute of Chinese Materia Medica, and Luo Zhonghan, Chengdu University of Traditional Chinese Medicine;
12. Simple *Qinghao* preparations in treating malaria, by Lu Ziyi, Gaoyou County Ministry of Health, Jiangxu Province.

***

B. Credit Assignments

The discovery of *qinghaosu* was the result of the accumulated efforts of countless research units, scientists, and administrative personnel. The "evaluation process" validated the results of the *qinghaosu* research and also confirmed the contribution in different ways of all participating research units. This was necessary because the prevailing social and political imperatives dominant in China at the time required that each research and administrative unit involved in the *qinghaosu* development project was given appropriate credit for its contribution. Naturally, the units had difficulty in deciding among themselves the importance of their respective discoveries. Following a heated debate moderated by the chief organizer of the meeting, a consensus was reached that six principal research units would be recognized in the following order:

1. Institute of Chinese Traditional Medicine, Ministry of Health;
2. Shandong Institute of Combined Western and Chinese Traditional Medicine;

3. Yunnan Institute of Materia Medica;
4. Guangzhou College of Traditional Chinese Medicine;
5. Sichuan Institute of Traditional Chinese Materia Medica;
6. Jiangxu Gaoyou County Department of Health.

In addition to the principal research units, thirty-nine collaborating units were named, including units from the Institute of Biophysics and the Shanghai Institute of Organic Chemistry, both Chinese Academy of Sciences.

Details of this evaluation meeting on *Qinghaosu* research were published in the first issue of the *Chinese Journal of Medicine* in 1979.

***

C. Contributions of the Various Research Units in Chronological Order

From the earliest reports on the Chinese medicinal plant *Qinghao* to the successful completion of *qinghaosu* development, there was an ongoing debate, between units, regarding who discovered what first. The following account is based on the original records documenting the results in extracting *Qinghao* crystals and on the clinical trials performed by various units, in chronological order where possible.

***

C1. Qinghaosu (Huanghaosu, Huanghuahaosu) Extraction

December 1972: The Beijing Institute of Traditional Chinese Materia Medica extracted *Qinghao* crystals from *Qinghao* plants growing in Beijing. Experiment code "*qinghaosu* II" and later renamed "*qinghaosu*."

April 1973: The Yunnan Institute of Materia Medica made ether extract crystals from *huanghuahao* with large flower-heads growing in Kunming District. Experiment code *"Kuhao* crystals III" and later renamed *"huanghaosu."*

November 1973: The Shandong Institute of Chinese Traditional Medicine extracted seven kinds of crystals from *huanghuahao* growing in Taian District of Shandong. Crystals No.5 was named *"huanghuahaosu."*

***

C2. Clinical Trials

September to October 1973: In Hainan Changjiang District, using *"qinghaosu* II," the Beijing Institute of Traditional Chinese Materia Medica successfully treated three cases of vivax malaria, and had five failures in treating falciparum malaria (one responded briefly but relapsed on day six; two were totally ineffective; two terminated treatment because of cardiac toxicity).

May 1974: In Juye County Shandong Province, using *"huanghuahaosu,"* the Shandong Institute of Chinese Traditional Medicine and Shandong Institute of Parasitic Diseases treated nineteen cases of vivax malaria. The drug was effective against vivax malaria in northern China, with no apparent side effects. The drug was not tried on falciparum malaria in that area.

October 1974 to January 1975: In Yunnan Gengma County, using the *"huanghaosu"* (*Kuhao* crystals III) provided by the Yunnan Institute of Materia Medica, the clinical team of Guangzhou College of Traditional Chinese Medicine treated three cases of severe malaria, eleven falciparum malaria, and four cases of vivax malaria in an area known to have drug-resistant malaria, in a non-randomized comparison with chloroquine. During this time, the first comments on *huanghaosu* for treating falciparum malaria were made: it was fast acting, highly effective in the short term, and caused few side effects, but had a high recrudescence rate.

June 1975: After the Chengdu meeting with ten provincial, city, and district research units, the Beijing Institute of Traditional Chinese Materia Medica restarted clinical trials using the *qinghaosu* (*huanghaosu*) supplied by Wuhan Jianming Pharmaceutical Factory in Hubei Province. The extract was from *huanghuahao* growing in Sichuan Youyang County and extracted by a solvent-gas method developed by the Yunnan Institute. The clinical trials were carried out in Hainan, in Hubei Medical College, Wuhan Iron and Steel Corporation, and Henan, in collaboration with the local teams.

During this time, the chemical structure of *Qinghao* crystals was still unknown. With no reference data for quality standards and corresponding analytical controls, the *qinghaosu* of different places could not be compared. The question of whether the extracted crystals contained active antimalarial substance could be answered only by the results of clinical trials.

From the chronological records, it is apparent that timing of *Qinghao* crystal extraction by various units did not coincide with the chronological order of clinical trials. Therefore, after the preliminary clinical evaluation of *qinghaosu* was announced, there was a longstanding difference in opinion and ongoing debate as to who discovered *qinghaosu* first. On this subject, the national Project 523 head office, based on chronological records, maintained an objective and unbiased attitude towards all research units and their contributions to the project. This independent position taken by the head office was seen in many evaluation documents.

In February 1974, the four Project 523 initiators (Ministry of Health, Ministry of Chemical Industry, General Logistic Department of the Chinese People's Liberation Army, and the Chinese Academy of Sciences) jointly called a meeting in Beijing. The summary report of this meeting was distributed to administrative departments of all districts, all local Project 523 offices, and all research units involved in the project. The report states:

"Since 1971, the Institute of Chinese Materia Medica of the China Academy of Traditional Chinese Medicine, Shandong Institute of Parasitic Diseases, Shandong Institute of Chinese Traditional Medicine, and Yunnan Institute of Materia Medica, one after the other, obtained preliminary results from research experiments on *Qinghao*, including successful results from clinical studies."[21]

In October 1977, nearly a year before the *qinghaosu* evaluation meeting in November 1978, the national Project 523 head office submitted a summary report to the four Project 523 initiators, mentioned earlier, narrating the course of *qinghaosu* development. The order of institutes appearing in this report was still Beijing-Shandong-Yunnan. But the report also commented on certain of the research projects in ways that shed a different light on the order of discoveries. In describing the 1973 Beijing Institute of Traditional Chinese Materia Medica's clinical trial on eight cases using "*qinghaosu* II," the report stated, "Because the expected satisfactory results were not achieved, the research has been adversely affected for a while." On the clinical trials by the Guangzhou, Yunnan, and Shandong units, it said, "The clinical trials in Yunnan and Shandong with 44 malaria patients in 1974 further confirmed the antimalarial effects and other valuable characteristics of *qinghaosu (huanghaosu)*. This prompted the decision to undertake further *Qinghao* research."[16] This suggests that the Beijing trial was not influential in continuing *qinghaosu* development, but the later trials in Yunnan and Shandong were.

Lengthy discussions and debates about who discovered what and when were held at the November 1978 National *Qinghaosu* Evaluation Conference. The final "*Qinghaosu* Evaluation Report" concludes the following:

"In October 1971, the Institute of Chinese Materia Medica of the China Academy of Traditional Chinese Medicine found the active antimalarial component in the Chinese herb *Qinghao*

(*Huanghuahao*). In 1972 and 1973, the Institute of Chinese Materia Medica of the China Academy of Traditional Chinese Medicine, Shandong Institute of Chinese Traditional Medicine and Yunnan Institute of Materia Medica in turn isolated the active monomer *qinghaosu* (*huanghuahaosu, huanghaosu*). In 1974, Guangzhou College of Traditional Chinese Medicine and Yunnan Institute of Materia Medica successfully treated falciparum malaria and cerebral malaria with *qinghaosu*. From 1975, the "*Qinghao* Research Collaboration Group" was formed, comprising the China Academy of Traditional Chinese Medicine, research units from Shandong, Yunnan, Guangdong, Sichuan, Jiangsu, Hubei, Henan, Guangxi, Shanghai, the Chinese Academy of Sciences, and the research teams affiliated with the Chinese People's Liberation Army. This collaborative group carried out systematic, in-depth research into many aspects of *Qinghao*, such as plant resources, clinical studies, pharmacology, chemical structure, drug formulations, production procedures and techniques, and quality control for standardization of the finished product."[41]

The "*Qinghaosu* Evaluation Report" not only documented the chronological order of *Qinghao* research, but also evaluated whether the treatment was effective and confirmed the contribution of dozens of research units. Although opinions on the report differed, a general agreement was reached to accept it, since it expressed the overall positive spirit of this major multidisciplinary enterprise in achieving such a successful outcome after many years of collaborative research.

***

The Naming of Qinghaosu

Naming what we now know as *qinghaosu* was another important point of the *Qinghaosu* Evaluation Conference. In the nomenclature of traditional medicinal herbs, the plant

Huanghuahao and the plant *Qinghao* were both called "*Qinghao.*" *Qinghaosu* was extracted from the plant *Huanghuahao*, while the plant *Qinghao* did not, in fact, contain any antimalarial substance. So there were always two different opinions regarding the naming of the substance with antimalarial effect. One opinion suggested naming the active substance "*qinghaosu*" because "*Qinghao*" had been an important antimalarial plant in Chinese traditional medicine and the Beijing Institute of Traditional Chinese Materia Medica had called it "*qinghaosu*" right from the beginning of its discovery. The opposing opinion held that since *qinghaosu* was extracted from the plant *Huanghuahao*, and *Qinghao* itself did not contain the antimalarial substance, it was obvious for all practical purposes and from the academic point of view that it should be named "*huanghuahaosu*" or "*huanghaosu.*"

The debate over naming *Qinghao* extract was superficially an academic problem, but in reality it was another reflection of the ongoing argument on the order of importance of the research units involved. The Evaluation Conference could not come to a conclusion about naming and suggested that the research units concerned and experts in the fields discuss the problem among themselves after the meeting. The research results had already confirmed that only the plant *huanghuahao*, specifically the variant of *huanghuahao* with big flower heads, contained the antimalarial substance and that *Qinghao* did not contain an antimalarial component. The *China Pharmacopoeia* had already followed the customary Chinese medicinal herb usage because under the heading "Chinese medicinal herb *Qinghao* plant family" it listed only *huanghuahao*, i.e., *Artemisia annua L.* A decision was made to name the active antimalarial component "*qinghaosu*," keeping the traditional medicinal herb name, although scientifically it could be considered incorrect.

***

D. Six Institutions Awarded a Certificate for the Discovery of Qinghaosu

On December 28, 1978, the State Council announced a national award for the discovery of *qinghaosu*. The Ministry of Health, representing the national Project 523 leading group, nominated six institutions to the National Commission of Science and Technology for the "*Qinghaosu* Discovery Award." These six institutions were confirmed as the principal investigative research units in the 1978 *Qinghaosu* Evaluation Conference.

Members of the evaluating committee for The National Commission of Science and Technology's Award for Discovery felt that if *qinghaosu* was considered as a new drug discovery, its chemical structure should be established. If the chemical structure of a new drug was not known then it could not be confirmed as a "new drug." Therefore the committee suggested the institutions responsible for establishing the *qinghaosu* chemical structure (the Institute of Biophysics and the Shanghai Institute of Organic Chemistry, both affiliated with the Chinese Academy of Sciences) be added to the list of principal research units. These replaced the Sichuan Institute of Chinese Materia Medica and Jiangxu, Gaoyou County Department of Health, which made unpurified crude *Qinghao* tablets and simple preparations of crude *Qinghao*, and were considered as principal collaborative units, but not principal research units. In September 1979, with the agreement of the Ministry of Health, the National Committee of Science and Technology awarded the "Certificate for Discovery," signed by the Committee's Director Fang Yi, to the Academy of Chinese Traditional Medicine, Ministry of Health; Shandong Institute of Chinese Traditional Medicine; Yunnan Institute of Materia Medica; Institute of Biophysics of the Chinese Academy of Sciences; Shanghai Institute of Organic Chemistry of the Chinese Academy of Sciences; and Guangzhou College of Traditional Chinese Medicine. The discovery item was listed as a "new antimalarial drug, *qinghaosu*," winning

Second Prize and medal No.1005. When the certificate was awarded, the newspaper *People's Daily* announced to the nation and the outside world the birth of a "new antimalarial drug *qinghaosu*," a discovery of Chinese scientists.

In retrospect, the Jiangxu, Gaoyou County Department of Health and the Sichuan Institute of Materia Medica were major contributors to the successful development of *qinghaosu*. They were not, unfortunately, included in the list for the Discovery Award, but the important part they played in the project cannot be denied.

During the *Qinghao* and *qinghaosu* research period, under the direction of the Jiangxu Gaoyou County Department of Health, village doctors Gu Wenhai and others used medicinal herbs to treat malaria, based on rural traditions. They administered various simple preparations of *Qinghao* to a large number of patients, using multiple treatment protocols in well-organized and well-planned clinical trials. Their preliminary results confirmed that *Qinghao* simple preparations were effective in treating vivax malaria.

The Sichuan Institute of Chinese Materia Medica was assigned by Project 523 head office and the local 523 Sichuan district group as the research center for a major *qinghaosu* development effort. This was to use locally available material to extract better quality preparations with improved antimalarial activity. The Sichuan Institute mobilized more than twenty scientists and technical personnel from its chemistry and pharmacology departments and collaborated with forty military and civilian research units (medical, educational, scientific, and pharmaceutical) throughout Sichuan Province.

The research was carried out systematically, from the investigation into plant material resources, chemical extraction of active material, pharmacology and toxicology studies, production procedures for quality control, and clinical trials. The final product obtained was "*Qinghao*" tablets, a crude preparation of *qinghaosu*. In clinical trials in 590 cases of falciparum and vivax malaria, all patients were cured.[42] The pills were fast-acting with a high short-term efficacy and few side effects, and they were easy to make at a low cost from an abundance of readily available raw material. In the discussion at the *Qinghao* Research Evaluation Conference organized by the national Project 523 leading group, "*Qinghao* tablets" were evaluated as a single item and successfully passed the evaluation process. In the collaborative *Qinghao* research project, the Sichuan Institute of Chinese Materia Medica had successfully developed a production procedure using ethanol. This provided not only 600 grams of *qinghaosu* to the Beijing Institute of Traditional Chinese Materia Medica, but also sufficient amounts to other units for scientific experiments and clinical trials, and to the military on active duty in the field.

In addition to the six research units for *qinghaosu* and the twelve research units for unpurified, crude, simple *Qinghao* preparations, thirty-seven collaborative research units were recognized. These collaborative units carried out many clinical trials, resource material investigations, and established quality standards in *Qinghao* and *qinghaosu* research. These scientists and technological personnel followed the directives of their superiors, completed their assigned missions, and worked quietly but efficiently in the background. These unnamed researchers deserve appreciation for their contributions and are to be highly commended for their dedication.

The success of *qinghaosu* research was the result of a collective effort that reflects well on the nation's scientists and technologists. Although six institutions received an award and made original individual discoveries, they could not have achieved such success on their own. In addition to their important contribution, manpower, scientific knowledge, technological input, equipment, and financial resources from throughout the country were necessary to complete the project. In the end, using modern scientific and technological methodologies, Project 523 isolated, chemically identified, purified, standardized, and clinically tested a new antimalarial from a traditional medicinal herb.

***

E. *Production for Military Use*

In the late 1970s, the military urgently needed *qinghaosu* for use at the country's border areas. In response, the State General Administration of Medicine, the Logistics Department of the People's Liberation Army, and the National Ministry of Health ordered the province and district administrative departments of Yunnan, Guangxi, Sichuan, and Guangdong Provinces to produce *qinghaosu*. The designated production units included the Yunnan Institute of Materia Medica and Kunming Pharmaceutical Factory, the Sichuan Institute of Chinese Materia Medica and Chungqing Pharmaceutical Factory No.8, Guangxi Guilin Aroma Chemical Factory, and Guangdong Hainan Pharmaceutical Factory.[43]

All provincial and district departments took this order seriously and immediately made operational arrangements. These units were the earliest research institutes and pharmaceutical factories to attempt *qinghaosu* extraction and to test production capabilities. With the involvement of local Project 523 offices, the research institutes and production units decided on a reasonable production procedure, used all their production potential, cleared stored material, purchased *Huanghuahao*, and immediately proceeded to produce *qinghaosu*. The Sichuan Institute of Chinese Materia Medica used more than one million *Qinghao* tablets left over from previous research studies and clinical trials as a source for *qinghaosu* extraction. With the collective efforts of all districts, the production task was quickly completed, supplying hundreds of thousands of *qinghaosu* oil injections. Subsequent orders for the production of 100 kilograms of *qinghaosu* were also completed on time by all units. This was a test for Project 523 research units and production units. All members and participating teams of the *Qinghao* and *qinghaosu* project from all areas of expertise, especially the production-line workers who stayed overtime to reach the production target, gave themselves over to fulfilling the military needs. People were proud and pleased to be able to participate in the research, preparation, and production of the new antimalarial *qinghaosu* to serve the nation's military needs.

This episode of the emergency production for combat readiness expedited promotion for the wider use of *qinghaosu*, and also motivated further research on the drug. To satisfy the urgent needs of *qinghaosu* within and outside China, the issue of a permanent site for the industrial production of *qinghaosu* was on the agenda for discussion by members of major government departments.

CHAPTER 6

Derivatives with Improved Effectiveness: Artemether, Artesunate, and Dihydro-artemisinin

A. Shanghai Institute of Materia Medica: Chemical Structure Modification

The 1978 *Qinghaosu* Research Evaluation Conference in Yangzhou City, Jiangxu Province, announced the discovery of a new Chinese antimalarial drug *qinghaosu*. In the course of *qinghaosu* research and development, this was only the first phase. The second phase concentrated on the development of a series of *qinghaosu* derivatives and new compounds in combination with *qinghaosu*, for possible use in other diseases.

The original idea was presented in 1975 by the national Project 523 office, when the task of determining *qinghaosu*'s structure was almost complete. It arranged a meeting with experts in the appropriate fields to discuss research into *qinghaosu* derivatives with new structures, mechanisms of action, and improved efficacy.

In order to find a new generation of antimalarial drugs that had a higher cure rate than *qinghaosu*, had a lower recrudescent rate, and was more stable and easier to use, all experts agreed that one should first modify *qinghaosu*'s chemical structure to increase its solubility and bioavailability. In December 1975, as mentioned earlier, during the Shanghai Project 523 Synthetic Drug Evaluation Conference, the leader of the national Project

523 head office met with the department head and experts of the synthesis laboratory of the Shanghai Institute of Materia Medica to discuss modifying *qinghaosu*'s chemical structure. All agreed that this was the correct approach to improve the effectiveness of *qinghaosu*.

Since 1967, for Project 523, the Shanghai Institute of Materia Medica had been responsible for research into synthetic antimalarial drugs and Chinese antimalarial medicinal herbs. It was one of the main research units of Project 523. The institution had done excellent work in modifying the chemical structure of the synthetic drug β-dichroine, and in research into the antimalarial effects of a Chinese medicinal herb, *Herba agrimoniae*. Pyrozoline, derivative No. 56 of β-dichroine through chemical structure modification, had better antimalarial effects and less vomiting as a side effect than β-dichroine. But compared to other antimalarial drugs from Project 523, pyrozoline was less effective and therefore not used as an antimalarial drug, but had been developed into a new antiarrhythmic agent. Regarding the Chinese herb *Herba agrimoniae*, its antimalarial effects were confirmed by laboratory experiments and clinical trials. Its active monomer agrimophol was isolated and its chemical structure established.

In February 1976, the national Project 523 leading group assigned the task of modifying *qinghaosu*'s chemical structure to find new derivatives to the Shanghai Institute of Materia Medica. The Institute immediately organized researchers of the chemistry, phytochemistry, and pharmacology laboratories to work on the relationship between the chemical structure and *qinghaosu*'s antimalarial effect, and to develop derivatives.

Qinghaosu is a sesquiterpene lactone containing a peroxide group. The action of this peroxide group became the focus of chemical synthesis research. In the chemical synthesis of *qinghaosu* carried out earlier by the Shanghai Institute of Organic Chemistry, the chemical reactions produced various unknown compounds. The Shanghai Institute of Materia Medica discovered that the majority of these chemical compounds had lost the peroxide group or the parent nucleus of *qinghaosu*.

The only exception was that at a temperature around 5°C, when reduced by sodium borohydride, *qinghaosu* became dihydro-*qinghaosu* (dihydro-artemisinin), keeping its parent nucleus and its peroxide group. With this information, the Shanghai Institute of Materia Medica synthesized dihydro-artemisinin and deoxy-artemisinin, which had lost the peroxide group. Both were tested on *Plasmodium berghei* by the pharmacology department. The results indicated that the deoxy-artemisinin had no antimalarial effects, whereas dihydro-artemisinin containing the peroxide group had better antimalarial effects than *qinghaosu* (artemisinin). This was a very important experimental result and a very important discovery. It proved that the peroxide group was needed for antimalarial activity. If the peroxide group was effective against malaria, would other peroxides also be effective? A few simple peroxides were then synthesized (i.e., those containing one or two peroxide groups). The pharmacology department again performed experiments with *Plasmodium berghei* using these synthesized peroxides and ascaridol, a crude natural monoterpene peroxide. But the antimalarial effects of these chemical compounds were less than ideal. Thus it was assumed that in creating an effective *qinghaosu* derivative, it was necessary to keep both the peroxide group and the *qinghaosu* parent nucleus. The *qinghaosu* structure modification experimented on by the botanical chemistry department also showed that when the *qinghaosu* molecule underwent too much modification, it lost the antimalarial effects. Meanwhile they also studied *qinghaosu* metabolism in the human body, hoping to find clues to active substances of *qinghaosu* metabolism. However, these metabolites also had no antimalarial effects.

Although dihydro-artemisinin had better antimalarial effects than *qinghaosu*, its solubility had not been improved and it was even less stable than *qinghaosu*. On this basis, Li Ying and others of the chemical synthesis group designed and synthesized three kinds of dihydro-artemisinin derivatives: ether, carboxylic acid ester, and carbonic ester derivatives.

At the Project 523 "Combined Chinese Traditional and Western Medicine for Malaria Treatment and Prevention Conference" in April 1977, in Nanning, Guangxi Province, and again in June of the same year at the Project 523 "Synthetic Drug Specialty Group Meeting" in Shanghai, Yu Peilin, Ge Yuanzhun, and Qu Zhixiang from the Shanghai Institute of Materia Medica reported their research protocol. This cited various ways of creating artemisinin (*qinghaosu*) derivatives and made other suggestions (including a preliminary protocol for the water-soluble esters of artemisinin). They also presented the preliminary screening results of more than twenty artemisinin (*qinghaosu*) derivatives they had created. Gu Haoming tested the antimalarial activity of these derivatives and found dihydro-artemisinin was twice as effective as artemisinin (*qinghaosu*); SM224 (methyl ether) was six times as effective and the highest activity; SM227 (ether) was three times as effective as artemisinin (*qinghaosu*). Since then, in subsequent research, many derivatives with antimalarial activity ten times that of artemisinin (*qinghaosu*) have been found.

In November 1977, after discussions between the experts, it was decided to choose from each of the three kinds of artemisinin (*qinghaosu*) derivatives one compound that had the most antimalarial activity, and to test it on large animals for antimalarial activity and toxicity. The dose of these chemical compounds needed for the study was relatively high. In experiments by the phytochemistry group a few months earlier, they had discovered that direct acidification of the methanol reacting solvent obtained in the process of making dihydro-artemisinin would also produce SM224. So they used this method of production, markedly increasing the speed of making SM224. The synthesis chemistry group had also synthesized derivatives SM108 and SM242. Finally, based on the results of toxicology studies, stability, solubility, and the cost for production, SM224 was chosen as the candidate for testing in large animals. Following this decision, the Shanghai Institute of Materia Medica systematically experimented with SM224 in terms of its pharmacology, toxicology, drug absorption, distribution, excretion, pharmacokinetics, toxicity in pregnancy, teratogenicity, and *in vivo* drug metabolism.

To be ready for clinical trials during the malaria season of the coming year and under an arrangement with the Shanghai Project 523 local office, the Yunnan Institute of Materia Medica, the Sichuan Institute of Chinese Materia Medica, and Guangxi Guilin Aroma Chemical Factory provided the *qinghaosu* (artemisinin) for making derivatives. SM224 was very soluble in oil and easy to make into an oil injection form. Jiang Yishan and others from Shanghai Pharmaceutical Factory No.10 were responsible for preparing the oil injections of SM224 (artemether). Wang Zhongshan and others from the Shanghai Institute for Drug Control established the method for quantitative analysis.

***

B. Artemether Clinical Trials

In 1978 the national Project 523 head office arranged for the first clinical trial of the artemisinin (qinghaosu) derivative, artemether (oil injection form of SM224), at Hainan, carried out by the 523 clinical group of Guangzhou College of Traditional Chinese Medicine. The researcher from the Shanghai Institute of Materia Medica personally delivered the drugs for the clinical trial to the Hainan investigators and participated in the clinical study. To demonstrate the authenticity of the historical record, the following is an excerpt from the original report of the first clinical trial of SM224 (artemether):

"Time of the clinical trial: July to September 1978, at the peak of the Hainan malaria endemic season.

Site of clinical trial: Dongfang County People's Hospital, in a drug-resistant falciparum malaria endemic area.

Clinical researchers: Li Guoqiao, 523 Clinical Group of Guangzhou College of Traditional Chinese Medicine; and Guo Xingbo, Department of Combined Traditional Chinese Medicine and Western Medicine, Dongfang County People's Hospital.

Patients entered into the study: total 17 (1 with combined hemolytic and cerebral malaria; 14 falciparum malaria; 2 vivax malaria). 6 patients had taken chloroquine prior to admission, 2 of them had completed 1 treatment course.

Treatment protocol: 2 groups, total dose 240 mg (80 mg one daily injection for 3 days) versus 480 mg (160 mg one daily injection for 3 days)

Evaluation of SM224:

1. All 17 malaria patients clinically cured. Judging from the parasite clearance time and the fever normalization time, both groups had a faster response than 1200 mg of artemisinin as an aqueous injection. SM224 was easy to use;

2. Six of the 17 patients had taken chloroquine 10 days prior to SM224 injection, and 2 of these 6 had completed 1 chloroquine treatment course, but symptoms were worsening. After SM224 injections, treatment effects were satisfactory. Preliminarily results confirm that artemether is effective in treating chloroquine-resistant falciparum malaria;

3. Four falciparum malaria patients in SM224 240 mg group all had recrudescence within 1 month; 9 of 11 falciparum malaria patients in the 480 mg group had 1 month follow-up, 3 did not show recrudescence on day 20. Initial impression: SM224 total dose 480 mg was appropriate for treating falciparum malaria."

The foregoing was the evaluation for the first clinical use of artemisinin derivative SM224. Although there were only 17 patients, it was demonstrated that this artemisinin derivative was possibly better than artemisinin in treating malaria. The results of the clinical trial were consistent with the results of experiment in animals by the Shanghai Institute of Materia Medica.

"A good start is halfway to success" as an old adage says. The success of the 1978 initial trial in Hainan was a great encouragement to all scientific and technological workers in the Shanghai Institute of Materia Medica, but they were eagerly awaiting the results of future large-scale clinical studies. Under the arrangement by the national Project 523 head office, Yunnan Kunming Pharmaceutical Factory was responsible for this future task. In early summer 1980, Zhu Dayuan and Yin Munlung of the Shanghai Institute of Materia Medica went to Yunnan Kunming Pharmaceutical Factory to try making larger amounts of dihydro-artemisinin with the one-step reaction method using potassium borohydride instead of sodium borohydride, a method developed by Chen Zhongliang and Yin Munglung. Chief engineer Wang Dianwu of Kunming Pharmaceutical Factory headed the test production of the artemether oil injection preparation.

From 1978 to 1980, nearly a three-year period, under arrangements by the national Project 523 head office, clinical trials with SM224 (artemether) were carried out using a standard protocol in the malaria endemic areas of Hainan, Yunnan, Guangxi, Henan, and Hubei. The number of patients included was 1,088, among them 829 falciparum malaria cases (including 99 confirmed chloroquine resistant cases and 56 cerebral malaria cases), and 259 vivax malaria cases. The clinical results from all test sites were similar. In the 829 falciparum malaria cases, the short-term cure rate was 100 percent, and of these, 457 cases receiving 600–640 mg had temperature normalization time and parasite clearance time of 24–48 hours. The drug's effect in lowering temperature and in clearing parasites was faster than with chloroquine and other antimalarial drugs. Follow-up of

345 cured patients showed that at one month the recrudescent rate was 7 percent, far lower than the rate with artemisinin. This intramuscular injectable preparation was very convenient for treating severe falciparum malaria patients, and, importantly, chloroquine resistant cases had a cure rate of 100 percent. The reports from all clinical trial sites commented that SM224 had the following advantageous characteristics: highly effective with a low dose, fast-acting with rapid temperature normalization and rapid parasite clearance, low toxicity with only mild side effects, and easy to use. The improved preparation was well accepted by patients, especially children, because there was no swelling or pain at the injection site, and nurses and medical professionals were quite comfortable administering the drug.

By this stage, artemether, as the first *qinghaosu* derivative, had successfully completed all the required tests for a new drug. In January 20–22, 1981, the national Project 523 leading group headed the Antimalarial Drug Artemether Evaluation Conference in Shanghai. Participants comprised fifty-six representatives of thirty-seven provincial, city, and military affiliated institutes and research units from Beijing, Guangdong, Yunnan, Shandong, Henan, and Sichuan. Also attending were leaders and representatives from the National Commission of Science and Technology, the State General Administration of Medicine, the Logistics Department of the People's Liberation Army, the Shanghai Commission of Science and Technology, and the Shanghai branch of the Chinese Academy of Sciences, Shanghai Project 523 local office. The meeting attendees were impressed with the research results of artemether and felt that artemether was an important advance over artemisinin, and suggested the departments and units concerned should work to facilitate favorable conditions for the industrial production of artemether as soon as possible. In December 1996, artemether received the National Commission of Science and Technology's "National Invention Award" third prize.

***

C. Artesunate and Dihydro-Artemisinin

In June 1977, the Project 523 specialty group for chemical synthesis called for a meeting in Shanghai. At the meeting, the Shanghai Institute of Materia Medica introduced the qinghaosu (artemisinin) chemical structure modification and data showing the relationship between the effectiveness and the structure of the derivatives. The group raised the idea of increasing the solubility of the water-soluble derivatives of artemisinin. After the meeting and with the knowledge that there was an abundant resource of *Qinghao* in Guangxi Province, the Guangxi Project 523 local office requested Guilin Pharmaceutical Factory to start research on artemisinin derivatives.[44] Liu Xu of the Guilin Pharmaceutical Factory was also in the 523 specialty group meeting and listened to the Shanghai Institute of Materia Medica's presentation on artemisinin derivatives. Liu Xu then made use of the *Qinghao* resource and the production conditions for the intermediates in the production process of the Guilin Pharmaceutical Factory, to create more than ten derivatives. GuangxiMedical College and the Guangxi Institute of Parasitic Diseases then evaluated the effectiveness, pharmacology, and toxicology of these substances. The two most outstanding ones were code No.804 and No.887. For these new derivatives, the Guangxi Project 523 district office created the Guangxi *Qinghaosu* Derivatives Research Collaboration Group. Guangxi Aroma Chemical Factory would supply the *qinghaosu*. The Guangxi Medical Institute would be responsible for determining chemical structures of the derivatives by elemental analysis, infrared spectrum, mass spectrometry, and nuclear-magnetic resonance techniques. And the Institute of Biophysics of the Chinese Academy of Sciences would confirm the results with x-ray diffraction methodology.

Derivative No.804 sodium powder injection (dissolved in sodium bicarbonate solvent) was used in clinical trials in October 1978 in Guangxi, Nanning County. There were twenty-four patients in this study (nine with falciparum malaria, fifteen with vivax malaria). Results showed that in treating malaria, No.804 sodium powder injection in a total dose of 300 mg had a high short-term cure rate, was fast-acting, and had low toxicity, but had a high early recrudescent rate.

Derivative No.804 was a hemisuccinate of artemisinin, named artesunate. In March 1983, WHO Scientific Working Group on Chemotherapy (WHO SWG-CHEMAL) suggested prioritizing the development of artesunate for treating cerebral malaria, and also took an interest in the process of making artesunate preparations and initiating quality controls. Following the suggestion of WHO, the *Qinghaosu* Directional Committee arranged for the following institutes to undertake this development: the Shanghai Institute of the Pharmaceutical Industry, the Academy of Military Medical Science, the Institute of Materia Medica of the Chinese Academy of Medical Science, the Shanghai Institute of Materia Medica, the Beijing Institute of Traditional Chinese Materia Medica, Guangzhou College of Traditional Chinese Medicine, Guilin Pharmaceutical Factories No.1 and No.2, Guangxi Medical College, Guangxi University of Traditional Chinese Medicine, and the Guangxi Institute of Parasitic Diseases. All experiments and studies needed to be repeated according to WHO's standards, from refining of the raw material, to improvement of the manufacturing process, to product quality control, management of a germ-free workplace for packaging injection powder into vials or ampoules, toxicology studies, and clinical trials. On improving the manufacturing techniques and establishing methods of measuring artesunate blood levels, the Shanghai Institute of Pharmaceutical Industry and Beijing Institute of Materia Medica respectively had already completed important studies.

To repeat all the research on artesunate, the *Qinghaosu* Directional Committee unified the scientific research resources and manpower, unified research planning and arrangement, and provided financial support. In April 1987, the *Qinghaosu*

Directional Committee submitted the results for a new drug approval. From April to June 1987, the National Ministry of Health issued a "New Drug Certificate" to the eight following collaborative research units: Guilin Pharmaceutical Factory, Guangxi Medical College, the Guangxi Institute of Parasitic Diseases, Guangzhou College of Traditional Chinese Medicine, the Shanghai Institute of the Pharmaceutical Industry, the Institute of Microbiology and Epidemiology of the Academy of Military Medical Science, the Institute of Materia Medica of Chinese Academy of Medical Science, and the Institute of Chinese Materia Medica of the China Academy of Traditional Chinese Medicine. Certificate No. 87, registration No.X-01, was the registered drug name "artesunate." The injection preparation of artesunate was a joint effort of the Shanghai Institute of the Pharmaceutical Industry, Guilin Pharmaceutical Factory No.2, and Guangzhou College of Traditional Chinese Medicine. These three units shared a separate "New Drug Certificate" from the National Ministry of Health.

Artesunate was available in IV and IM forms, therefore adding more choices to the preparations of artemisinin and its derivatives for treating malaria and consequently increasing their clinical usage. After more than twenty years of commercial production and clinical use, artesunate has now become the first drug of choice for treating cerebral malaria.

The success in developing artesunate again demonstrated the value and importance of extensive collaborative efforts throughout the nation. Complying with the WHO requirements and standards for development of a new drug during the five years from 1982 to 1987, the *Qinghaosu* Directional Committee organized and financed a better and more efficient scientific and technological workforce. Later, with international collaboration, all the research requirements were completed for a successful development of this new drug. The record shows once more that the development of artesunate was not the product of one department or one individual research unit. In a medical journal, it is stated that "Artesunate's discovery, research, production and its registration and promotion in more than 30 counties in Asia, Africa and Latin America was completed by a single unit."[45] This is not correct.

On July 1, 1985, the National Ministry of Health published "Regulations for Evaluating New Drugs," with many added instructions. With the *Qinghaosu* Directional Committee's support, the collaborative units concerned complied with the new requirements cited in this publication, and carried out phase I and phase II clinical studies according to the necessary standards. Artesunate and artemether were studied under these new drug evaluation criteria as dictated by the "Regulations for Evaluating New Drugs," and the earlier results were confirmed so that they became representative examples of the nation's new drug evaluation process at the time.

***

After the dissolution of Project 523 offices in March 1981, the Beijing Institute of Traditional Chinese Materia Medica carried out the development of dihydro-artemisinin as a new antimalarial drug.

Dihydro-artemisinin: Earlier in 1975 to 1977, in establishing the chemical structure of artemisinin, and in the process of proving artemisinin was a peroxide, the Shanghai Institute of Organic Chemistry experimented with hydrogenation and deoxidization. It confirmed that the lactone of artemisinin could be reduced by sodium borohydride or potassium borohydride but still kept its peroxide group. The reduced substance was called reduced *qinghaosu* or reduced artemisinin and later named dihydro-*qinghaosu* or dihydro-artemisinin. It was an important intermediate substance for making artemisinin derivatives. In 1978 in studying the antimalarial effects of artemisinin derivatives, the Shanghai Institute of Materia Medica found that dihydro-artemisinin had better antimalarial effects than artemisinin. But it was not chosen for further study because of its chemical instability, low solubility, and less antimalarial effects than other derivatives.

In 1990 the Beijing Institute of Traditional Chinese Materia Medica revised the interest in dihydro-artemisinin, and invited the Chinese Academy of Medical Science and the Institute of Microbiology and Epidemiology of the Academy of Military Medical Science to reevaluate its antimalarial mechanism, pharmacology, and drug safety. Guangzhou College of Traditional Chinese Medicine carried out clinical trials and demonstrated that it had good antimalarial effects. All studies for dihydro-artemisinin were completed in 1992, and it passed the new drug registration process, obtaining a "New Drug Certificate." It was manufactured by Beijing Pharmaceutical Factory No.6, and promotion for sales into the international markets was done by the Beijing Cotec New Technology Corporation. The reevaluation of dihydro-artemisinin added a new member into the family of antimalarial artemisinin derivatives, thereby providing another drug to the armamentarium for treating malaria. This was also an important start for the future development of compounds to be used in combination with dihydro-artemisinin.

It should be noted that a claim has been made on the Internet and in newspapers and magazines that a person created dihydro-artemisinin in 1973.[46] It was widely known that the chemical structure of *qinghaosu* (artemisinin) was established after 1975. If someone could "create dihydro-artemisinin" before artemisinin's chemical structure was known, then there was no need to mobilize so much manpower, spend so much money, and take such a long time using highly technical equipment to do research on *qinghaosu*'s parent nucleus and the structure of its various groups. Without first determining the chemical structure of *qinghaosu*, how would one know the chemical structure of its derivative dihydro-artemisinin? The claim, therefore, of having created dihydro-artemisinin before the chemical structure was known has no scientific basis or support.

CHAPTER 7

International Considerations

A. WHO Involvement

Scientific and technological research activities in China became more open and progressive when the Cultural Revolution finally came to an end in the mid-to-late 1970s. A wide variety of scientific conferences were held, and scientific periodicals resumed publication. Since 1967 ten years of Project 523's research into the search for new antimalarial drugs had accumulated a vast amount of scientific data. During that time, when there was no patent law or official regulations to protect the rights to the products of one's research, publishing one's findings was the only way for scientific and technological workers to gain recognition for their work. The publication on *qinghaosu*'s chemical structure appeared in the 1977 *Chinese Science Bulletin* with "*Qinghaosu* Chemical Structure Research Collaborative Group" as the author. In the English edition of the 1979 *Chinese Medical Journal*, an article on the antimalarial effects of *qinghaosu* was published, giving experimental data and clinical trial results. The author was "*Qinghaosu* Research Collaborative Group." Soon after, more *qinghaosu* articles appeared in scientific periodicals, and these were written by scientists and researchers under their own names.

After the results of the new Chinese antimalarial drug *qinghaosu* were publicized in early 1979, Li Ying from the Chemical Synthesis Laboratory of the Shanghai Institute of

Materia Medica wrote a summary of the completed *qinghaosu* derivatives research activities. She submitted it to the *Chinese Science Bulletin*, which published it in the latter half of 1979. In 1980 in the first issue of the *Acta Pharmacologica Sinica*, Gu Haoming and others from the Department of Pharmacology of the same institute as Li Ying published an article called "Effects of Artemisinin (*Qinghaosu*) Derivatives on Chloroquine-Resistant Strains of *Plasmodium berghei.*" They reported the antimalarial effects against *Plasmodium berghei* [SD90] of twenty-five artemisinin derivatives. In the same year, Liu Xu from Guangxi Guilin Pharmaceutical Factory also reported the synthesis of artesunate in the *Pharmaceutical Bulletin*. All these articles drew the attention of the World Health Organization (WHO) and other scientific organizations in countries outside China, as they followed the data being published on our new antimalarial drugs.

On December 5, 1980, the WHO's director general, Dr. Mahler, wrote to Minister Qian Xinzhong of the National Ministry of Health, stating that the spread of multidrug-resistant strains of *Plasmodium falciparum* had become a major threat in the world. He stated that the WHO Scientific Working Group for Malaria Chemotherapy (SWG-CHEMAL) was keen to organize an "Antimalarial Drugs *Qinghaosu* and Its Derivatives" conference in China, to discuss the possibility of helping China to further develop this new class of antimalarial drugs. After the National Ministry of Health accepted this request for a meeting, a five-year-long collaboration began between China and WHO.

Following a year of preparation, the conference was held in Beijing during October 6–10, 1981. It was the fourth meeting of the WHO SWG-CHEMAL group and the first to be held outside Geneva, Switzerland, the headquarters of WHO.

To prepare for this conference, all units involved in the drug research program combined efforts to resolve the weak aspects in the research on *qinghaosu*, and to supplement the already available data. These problem areas were in pharmacology, toxicology, and drug metabolism. Units involved included the Academy of Military Medical Science, the Pharmacology Institute of the Chinese Academy of Medical Science, the Shanghai Institute of Materia Medica of the Chinese Academy of Sciences, and the Beijing Institute of Traditional Chinese Materia Medica.

The purpose of the meeting was to evaluate and plan further development of artemisinin (*qinghaosu*) and the derivatives. Attending were members of WHO SWG-CHEMAL, Professor and Director N. Anand of the India Central Drug Research Institute, Director Dr. Brossi from the Department of Drug Chemistry at the US National Institutes of Health, Director Colonel C. J. Canfield of the Therapeutic Experimental Laboratory Walter Reed Army Institute of Research, Professor W. Peters, formerly director of the Department of Animal Research at the London School of Hygiene and Tropical Medicine, WHO officers W. H. Wernsdorfer and P.I. Trigg, and Dr. D. F. Clyde, consultant for malaria in Southeast Asia.

Participants from the Chinese side were leaders and representatives of the main administrative departments and research units, and experts in pharmacology and drug chemistry. They included Chen Haifeng, Gu Haoming, He Bin, Ji Ruyun, Ji Zhongpu, Jin Peihua, Li Guoqiao, Liang Xiaotian, Liu Erxiang, Liu Jingming, Liu Xirong, Liu Xu, Shen Jiaxiang, Song Zhenyu, Teng Xihe, Tu Youyou, Wang Pei, Wang Tongyin, Yang Lixin, Zhu Hai, Zhou Keding, Zhou Weishan, Zhou Tingchong, and Zhou Zhongming. The two chairmen of the conference were Professor Anand and Chen Haifeng, director of the Science and Technology Department, in the Ministry of Health.

Chinese representatives presented seven scientific papers covering the following: isolation and determination of the chemical structure of qinghaosu (artemisinin); chemistry and chemical synthesis of artemisinin and its derivatives; antimalarial effects; evaluation and preliminary research on mechanism of action; drug metabolism and pharmacokinetics; acute, subacute, and special toxicity studies; and the results of clinical trials. After each presentation, there were question and answer sessions and a thorough discussion. Finally three discussion groups were formed to promote ideas and exchange information in the areas of chemistry, pharmacology and toxicology, and clinical research.

The conference was very successful, with the scientific reports creating a high level of interest and a positive reaction from the foreign participants. The conference summary report of WHO SWG-CHEMAL confirmed and evaluated highly the antimalarial effects of artemisinin and its derivatives. The report stated that "An ideal new drug should have a new chemical structure and a new mechanism of action to possibly delay the development of drug resistance, and therefore prolong the useful life of the new drug. Also there is a need for a tolerated, safe, and fast acting preparation for treating severe falciparum malaria, especially cerebral malaria. It is apparent that the Chinese antimalarial drug qinghaosu (artemisinin) has these main required characteristics, because it has a new chemical structure, its mechanism of action seems to differ from the current drugs in clearing the trophozoite stage of parasite development, and it is fast acting."

The conference chairman, Professor Anand, also stated that "The significance of the artemisinin discovery and the success of the artemisinin derivatives research lie in the drug's unique chemical structure, and its antimalarial mechanism of action, which is different from any of the known antimalarial drugs. This opens a new direction for the designing of future new antimalarial compounds. So it is not surprising to see that artemisinin types of compounds are effective against chloroquine-resistant strains of *Plasmodium* parasites."

In view of the rapid spread of chloroquine-resistant strains of *Plasmodium* in Southeast Asia, Western Pacific, South America, and East Africa, the meeting was keen for the Chinese scientific researchers to cooperate with WHO SWG-CHEMAL to further investigate and develop these types of compounds. The meeting requested China to strictly follow international standards to complete the necessary preclinical work quickly, and within two years if possible. This included improving the formulation of different drug preparations, and research into teratogenicity and any toxicity in early pregnancy, so that this new drug could be made available and used on a large scale as soon as possible.

Finally, the meeting agreed on a "Development Plan for Artemisinin and its Derivatives," suggesting research topics for the next two years and possibly further, and laid out a preliminary research plan for the cooperation between WHO SWG-CHEMAL and China. To strengthen the leadership and the smooth operation of the research plan, WHO suggested China set up a corresponding administrative organization to communicate with the WHO secretariat and to coordinate the carrying out of the research plan. WHO would provide support and finances according to an agreed upon document. WHO SWG-CHEMAL would publish the scientific reports and technological data presented in this meeting as a monograph.

Leaders, scientists, and technologists of the institutes involved in the meeting commented afterwards that the nation's medical and scientific achievements had been demonstrated, and the international exchange of knowledge was to be commended.

The meeting and discussions also showed us that in new drug research, especially in the technology of drug preparation and preclinical pharmacology and toxicology studies, we were far behind international standards and had some weak areas, and also that our knowledge was deficient in certain fields. Discovering these shortcomings was to our benefit since it would lead to improvement and progress. Many participants at the meeting commented that China had been isolated from the world for too long and was unaware of important advances in international standards. With a good research plan and in cooperation with WHO, it would be possible to quickly elevate our new drug research to an international level. Everyone had great expectations for future cooperation with WHO.

During the conference, the leaders of the National Ministry of Health and the National General Administration of Medicine attended several sessions. At the conclusion of the conference, Minister Qian Xinzhong of the National Ministry of Health met with members of WHO SWG-CHEMAL and other international participants.

***

B. Establishment of Qinghaosu Directional Committee

As a result of the changes occurring within China and in the world in general, a report was issued on August 27, 1980 by the following organizations: the National Ministry of Health; the National Commission of Science and Technology; the National General Administration of Medicine (previously a member of the national Project 523 leading group, which, during the Cultural Revolution, merged with the Ministry of Petroleum and Chemical Industries, and after the Cultural Revolution again became the National General Administration of Medicine); and the General Logistics Department of the Chinese People's Liberation Army. The report was sent to the State Council and the State Central Military Commission to suggest discontinuation of the national malaria research leading group. The report summarized the completion of Project 523 and the results of thirteen years of scientific research, and also mentioned the supply of drugs to help allies in military combat. Following China's new direction in adjusting and changing the economic situation, the report suggested the elimination of the national and district Project 523 leading groups and offices. With the approval of the State Council from 1981, malaria research continued to be the major project for national scientific research in medicine and health at all levels. It was to be incorporated into routine research planning from the ministry level to the district level.

In March 1981, a "summit meeting" in Beijing was called jointly by the Ministry of Health, the Ministry of Chemical Industries, the General Logistics Department of the Chinese People's Liberation Army, the National Commission on Science and Technology, and the State General Administration of Medicine. The purpose was to bring together all group leaders and office directors of all district malaria research leading groups and offices. Deputy Minister Huang Shuze of the National Ministry of Health, representing the national Project 523 leading group, summarized the achievements of the project. Minister Qian Xinzhong, Deputy Minister Zhang Ruguang of the General Logistics Department of the Chinese People's Liberation Army, and members responsible for Project 523 from the National Commission of Science and Technology and the State General Administration of Medicine also spoke at the meeting. Diplomas were given to 110 research units, and certificates of honor were given to 105 long-term leaders and coordinators of the project. The meeting also made suggestions for the next step in the ongoing scientific research.

After termination of Project 523 and the elimination of its administrative structure, and following WHO's suggestion, the "*Qinghaosu* Directional Committee" was established on March 20, 1982. It comprised the National Ministry of Health and the State General Administration of Medicine. Director Chen Haifeng of the Department of Science and Technology of the National Ministry of Health was the head of the committee; Wang Pei, deputy dean of the Institute of Traditional Chinese Medicine, and engineer She Deyi from the Science and Education Department of the State General Administration of Medicine were appointed as deputy heads of the committee. Leaders and experts from the principal research units involved were appointed as members and consultants to the committee. To strengthen the communication links with WHO and other overseas organizations, Jin Yunhua and general engineer Shen Jiaxiang from the State General Administration of Medicine were invited as consultants to the committee. They facilitated cooperation between China and the international community in relation to activities involving artemisinin and its derivatives. Under the *Qinghaosu* Directional Committee, there were four specialty groups: chemistry, pharmacology, clinical pharmacology, and drug formulation. After termination of the national Project 523 leading group and head office, the *Qinghaosu* Directional Committee took over the organization of the nation's research and development of artemisinin and its derivatives. The committee made plans and designs for artemisinin production, organized the manufacturing of the products, managed product quality, promoted international exchange, encouraged collaboration with WHO and overseas pharmaceutical industries, and introduced into China advanced technology and a regulatory system for new drugs. In this new era of development, the committee took on a new management model of a unified, centralized purpose. It also acted as a coordinator in the relationship between Chinese units and international organizations. The establishment of the committee played an important role in the history of the development of artemisinin and its derivatives.

***

C. Good Manufacturing Practice (GMP) Problems

Following a series of discussions and preparatory procedures, WHO SWG-CHEMAL sent Secretary Dr.Trigg and scientific consultants Drs. M. Heiffer and C.C. Lee to visit China during February 1–14, 1982. They visited research units and pharmaceutical factories in Beijing, Shanghai, Guilin, and Guangzhou. At the end of the visit, Chen Haifeng discussed with the WHO representatives problems relating to technical requirements and financial support for the collaborative research

In 1968 Zhang Jianfang, Deputy Director of the National Malaria Prevention and Treatment Research Leading Group Office (National Project 523 Head Office) visited Project 523 work site at a minority group village in Hainan Island, China.

In 1968 Li Guoqiao (left) treating a malaria patient with acupuncture in a Li minority village in Hainan Island, China.

In 1974, Wang Zhanfa, Li Guoqiao, and Wang Zicai, from left to right, of the Guangzhou Project 523 clinical research team, working in a malaria endemic rural area in Yunnan Province, China.

May 2012, reunion of the above three colleagues in Guangzhou University of Traditional Chinese Medicine, (left to right: Wang Zhanfa, Li Guoqiao, and Wang Zicai).

Participants of the 4th meeting on Qinghaosu between WHO's SWG-CHEMAL and China, Beijing, China, October 1981.

Processing of qinghaosu (artemisinin): from fresh and dried qinghao plants, to dried leaf powder, crude preparation, crude extract, to qinghaosu crystals in capsules.

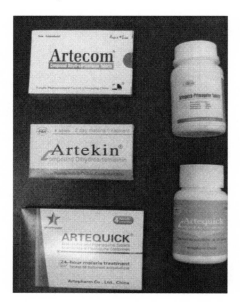

Various qinghaosu (artemisinin) based antimalarial pharmaceutical formulations.

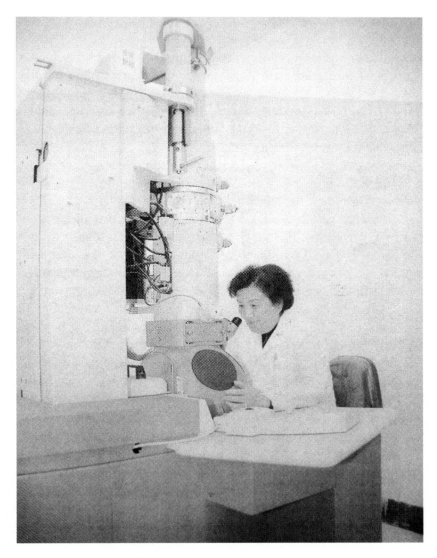

1976, Li Zelin of the Institute of Chinese Materia Medica, China Academy of Chinese Traditional Medicine examined the effects of qinghaosu on malaria parasites with electron microscopy.

in China over the next two years. A collaboration agreement was reached that stated if China was to be able to register and market their drugs internationally, certain criteria had to be met—namely, that within two years China had to complete quality control for pharmaceutical preparations of oral formulations of artemisinin and injectable forms of artemether and artesunate. Also full data on preclinical pharmacology and toxicology experiments, complete phase I, II, and III clinical studies, and all such work had to follow international standards. WHO would provide training for technical staff, supply equipment, and the necessary animals for the required studies.

In March 1982, WHO SWG-CHEMAL had a meeting in Geneva, reviewing the agreement signed with China a month earlier. SWG-CHEMAL considered artesunate for cerebral malaria the priority project but also expressed concern regarding the manufacturing of the drug. SWG-CHEMAL suggested the US Federal Drug Administration (FDA) scientific and technical members visit China to observe the drug manufacturing process and the management of the pharmaceutical factories.

To prepare for the GMP inspection by WHO and the FDA, the *Qinghaosu* Directional Committee provided funding to make the necessary technical modifications to the physical production facilities. The two plants involved were Guilin Pharmaceutical Factory No.2 and Kunming Pharmaceutical Factory, the former for manufacturing artesunate powder for injection, and the latter for an artemether oil injection preparation. The necessary air purifiers and air conditioners were added. The Shanghai Institute of Pharmaceutical Industry and the Academy of Military Medical Science sent experts to the factories. They helped train the key technological staff and gave instructions for improving the production process and prepared technical operational manuals. For the laboratory research units, the *Qinghaosu* Directional Committee encouraged improvement in the quality of the experimental work to meet the requirements of Good Laboratory Practice Guidelines (GLP). For clinical studies, the Department of Pharmacology of Beijing Institute of Traditional Chinese Materia Medica worked closely with the Department of 523 Clinical Study Group of Guangzhou College of Traditional Chinese Medicine, to carry out a phase I study with artemisinin suppositories following standard international methodology.

According to international regulations, before registration of a new drug, there must be an on-site inspection by officials from a legally recognized public institute. These officials examine the production conditions and management of the manufacturing factories, a process referred to as the GMP inspection. The inspectors then submit an evaluation report. At the suggestion of WHO and with the approval of the Chinese government, during September 21–28, 1982, Inspector Tetzlaff from the US Food and Drug Administration, accompanied by Secretary Dr. Trigg of WHO SWG-CHEMAL, carried out GMP inspections in Kunming Pharmaceutical Factory and Guilin Pharmaceutical Factories No.1 and No.2. The inspection emphasized a germ-free production workplace for manufacturing artesunate powder for injection in Guilin Pharmaceutical Factory No.2. The inspection process started with the inspector describing the GMP requirements for manufacturing a drug, and then he inspected the site item by item as listed in the requirements and examined the original records of production. The inspector then asked questions, had discussions with the factory's technological staff, and finally wrote the inspection reports.

The GMP inspection comments on the germ-free workplace for manufacturing artesunate injection in Guilin Pharmaceutical Factory No.2 were rather critical. They stated that the manufacturing process lacked strict regulatory management controls, the technological procedures lacked a scientific basis, especially regarding the disinfection procedure for a germ-free environment, and methods for testing disinfection were inadequate; the physical design of the factory and instrument maintenance were far from ideal. The conclusions: Based on the foregoing criteria, the workplace for producing artesunate IV injection in Guilin Pharmaceutical Factory No.2 did not meet GMP requirements and its products could not be used in clinical trials outside China.

Regarding the GMP inspection of Kunming Pharmaceutical Factory, the FDA inspector did not write a detailed report, but indicated similar problems as for Guilin Factory No.2. In order to find a pharmaceutical factory in China that fulfilled GMP requirements for the quick production of artesunate injection for clinical trials outside China, the FDA inspector examined Shanghai Xinyi Pharmaceutical Factory, which was considered the best-equipped in China. This factory also failed the inspection.

Although in several pharmaceutical factories the workplace for the end products of the manufacturing process, namely the pharmaceutical formulations, did not pass GMP inspection by the FDA, for the factory staff and the Chinese officials accompanying the inspector, this was an opportunity to learn about GMP. It was clearly necessary to not only improve factory facilities but also upgrade management behavior and the training of scientific and technological personnel in order to elevate production standards.

The FDA's GMP Inspector Tetzlaff explained that strict regulations and scientific production management methods were to ensure that the factory's drug products met all the quality control standards; only then could drugs be considered safe and effective. He summarized the essence of GMP in two words: "writing" and "signature."This meant that from the first step of drug production (mixing combinations of chemicals or adding material) to the drugs leaving the factory, each operational step or procedure and technical requirement need to have a written protocol or manual. In this way, the technical staff knows what they should or should not do. Also, in the operation, each procedure or technical step must be checked and recorded in writing and signed by the technical staff performing the procedure or the supervisor who checks the work, so that full responsibility is taken by both parties.

In Guilin Pharmaceutical Factory No.2, Inspector Tetzlaff had a few demonstrations that greatly impressed the staff. A row of glasses were placed on each side of the laboratory bench in the germ-free production workplace; one row was sterile glasses, and glasses in the other row had been used, but none were labeled with stickers "sterile" or "used." He asked the staff to turn their head, then picked up two glasses from the counter and asked: "Who knows which glass is sterile?" Obviously no one had the answer, including the supervisor of the workplace. Inspector Tetzlaff very seriously said, "This is GMP. Things look simple, but they are not so in practice." One technical director of a pharmaceutical factory commented that in the past, established behaviors, habits, and personal experience had become accepted as if they were scientifically based, or had the authority of a regulation or the status of an instruction manual. This on-site GMP inspection by the FDA was beneficial, and it was realized that conforming to the established requirements was absolutely necessary. It was important not only because a factory should be properly managed but also because the health and safety of perhaps millions of people depended on GMP. The future of the pharmaceutical industry also depended on compliance with GMP.

***

D. Failure of Negotiations and Collaboration Terminated

Because the production facilities did not pass GMP inspection, a "red light" came on in respect to the collaboration between China and WHO. This affected not only the production of artesunate but also a whole series of preclinical research projects, including preparation, production engineering, stability and pharmacology, and toxicology research studies. The situation was serious! On September 30, 1982, Chen Haifeng, head of the Department of Science and Technology, Ministry of Health, and head of *Qinghaosu* Directional Committee, met with Dr. Trigg of WHO SWG-CHEMAL to discuss the next step in the collaboration. Dr. Trigg suggested two alternatives: 1) build a production facility in China for the production of artesunate for injection that complied with the GMP requirements; this could delay international registration of artesunate 2–3 years; 2) collaborate with pharmaceutical companies outside China to manufacture an artesunate preparation up to GMP standards. Furthermore, he suggested completing the preclinical pharmacology and toxicology studies required for international drug registration as soon as possible, while also building a production facility in China that complied with the GMP requirements for later production.

The *Qinghaosu* Directional Committee thoroughly considered the pros and cons of Dr. Trigg's suggestions and decided that early international registration was the best choice. On November 11, 1982, the committee wrote to WHO SWG-CHEMAL, agreeing to seek collaboration outside China to manufacture artesunate and, if possible, to do the preclinical pharmacology and toxicology studies, since there was concern that the inadequate GLP conditions of the Chinese laboratories might affect international registration. The committee wanted WHO to recommend suitable research units and to suggest an appropriate collaboration agreement.

On January 4, 1983, the committee received a reply from Dr. Trigg suggesting collaboration with the Walter Reed Army Institute of Research in developing artesunate. On February 14, Dr. Trigg went to the Walter Reed Army Institute of Research to discuss this collaboration with China. As a consequence of this meeting, Colonel Brown of International Health Affairs of the US Department of Defense and Dr. Heiffer of the Walter Reed Army Institute of Research planned to visit China on May 30, 1983, to discuss collaboration on artesunate research. Unfortunately, because of the short notice, the committee was unable to make the necessary preparations and wrote requesting that the trip should be postponed, and could the US potential collaborative group send their proposal in advance. On December 22, 1983, Dr. Trigg sent the draft "Artesunate Research Collaboration Agreement," and stated the agreement had been approved by WHO and US Department of Defense, and hoped that China could arrange a meeting between the three parties as soon as possible.

After receiving the official agreement on December 23, 1983, the new department head of the Department of Science and Technology of the Ministry of Health, and also the new head of the *Qinghaosu* Directional Committee, Xu Wenbo, called a meeting to discuss the agreement drafted by WHO and US Department of Defense. Because of possible cultural differences, the agreement was found to be unacceptable. The differences related to what constituted "mutual benefit" and "friendly collaboration" and was felt to be one-sided. The participants also voiced the opinion that the agreement was too restrictive on China and that many changes needed to be made.

On February 17, 1984, Director Lucas of WHO/TDR informed China of the coming visit on March 13, 1984, to discuss the "Artesunate Research Collaboration Agreement." On May 3, 1984, the Ministry of Health and the State Administration of Medicine jointly submitted the "Artesunate Research Collaboration Agreement" for approval by senior government officials.

Two years had passed since WHO suggested that China collaborate with the Walter Reed Army Institute of Research in September 1982, to the time when both parties agreed on the preliminary draft of the collaboration agreement in October 1984. During this period, countless letters with endless debates on the rights, benefits, and responsibilities of each party were exchanged, but there was never an actual meeting at which the three parties discussed the matter face-to-face.

***

In the early 1980s, countries outside China showed an interest in artemisinin. The Swiss pharmaceutical company Roche synthesized artemisinin in 1982. The Walter Reed Army Institute of Research extracted artemisinin from locally growing plants along the Potomac River in Washington, DC, and determined its physical and chemical characteristics. Some experts from India and England who attended the October 1981 WHO *Qinghaosu* and Its Derivatives meeting in Beijing started experiments in growing *Qinghao* plants, and others began artemisinin pharmacological studies. Director Lucas of WHO's Department of Tropical Diseases had warned us that "the subject of your research has the risk of being taken away by others." Of course, at that time he made this statement in order to urge China not to bargain but to go ahead with the US on the "Collaboration Agreement," but in reality it clearly meant, "You don't have secrets to keep any more." Dr. Lucas's statement was obviously correct. Under these circumstances, China had no choice but to ask WHO to use its influence on countries and organizations to request that they respect China's rights to this discovery and to better manage the relationship with China to facilitate future collaboration. Unfortunately there was little that could be done to protect China's rights to the discovery of *qinghaosu*. We had been unaware, having been isolated for so long, that once something is published it becomes public property if international patents have not been applied for and registered.

After two years of debates and arguments, the collaboration with the US on artesunate research, with WHO acting as mediator, was brought to an end with no agreement being possible.

***

E. Problems with Artemether and Arteether

Both artemether and arteether (an ether derivative of dihydro-artemisinin) could be considered sister compounds. Both were first synthesized in 1977 by researchers at the Shanghai Institute of Materia Medica. Their methods of synthesis, physical constants, and active antimalarial effects [SD90] were published in the *Chinese Science Bulletin* in 1979 and in the *Acta Pharmaceutica Sinica* in 1981. The antimalarial effects of artemether was double that of arteether. China therefore authorized registering artemether as a new drug. But in 1985 WHO decided to develop arteether. It is difficult to understand why WHO did not support the more effective and clinically tested artemether rather than develop the less effective arteether.

However, there is an explanation. In the first half of 1985, members of WHO's SWG-CHEMAL and Dr. A. Brossi, who was head of the Department of Drug Chemistry of the US National Institutes of Health, visited the Shanghai Institute of Materia Medica. Researchers at the institute presented the artemether research to the visitors. At the time, artemether was already being successfully manufactured in Kunming Pharmaceutical Factory, and had known characteristics of good antimalarial effects, low toxicity, chemical stability, and no cross-resistance with chloroquine. The visitors were very interested in the good chemical stability because the progress of their priority studies with artesunate in treating cerebral malaria since 1982 were less than satisfactory. But there were unfounded rumors from CHEMAL that artemether metabolized in the body to release methanol, thereby increasing its toxicity.

On April 6–7, 1986, after a year of preparation, a plan for the total development of arteether was decided upon at the Geneva CHEMAL meeting. In the report on the ether derivative of dihydro-artemisinin presented to the meeting, Dr. Brossi mentioned that the members of CHEMAL in 1985 had decided on arteether and not artemether as the target drug for development by CHEMAL. In discussing the reason for choosing the ether derivative, Dr. Brossi referred to published reports. These stated that although these two compounds had similar antimalarial effects, choosing the ether derivative would avoid the problem of *in vivo* methanol release by artemether, which could then change into formaldehyde and formic acid with subsequent toxic effects. The source of this information was allegedly from a personal communication with a Chinese scientist.

The decision to drop artemether collaborative research and to develop the ether derivative by itself, based solely on a personal communication, without clarifying with the Shanghai Institute of Materia Medica and without verifying the error with scientific experiments, was considered really troublesome. Many Chinese scientists could not understand how this could happen.

In reality, in the "Plan for Future Research and Development of Arteether" read at the CHEMAL meeting, half of the items concerned support for improvement and cultivation of *Qinghao* resources, increasing the productivity of artemisinin, and strengthening the research on techniques for converting artemisinin to arteether. Meanwhile as regards to long-term drug supplies, the report mentioned that besides China as an abundant source for *Qinghao,* researchers in US, Europe, and Asia had already extracted artemisinin from *Qinghao* plants. But commercial production of artemisinin outside China and large-scale conversion into arteether still needed to be worked on. It was not difficult to see that CHEMAL's decision to develop arteether was really CHEMAL's lack of confidence in Chinese collaboration. It was therefore no surprise that two years of negotiation came to nothing.

There may have been difficulties for CHEMAL in the research and development of the ether derivative arteether, because on February 26, 1987, WHO/TDR secretary Dr. Wernsdorfer and WHO/TDR official Mr.Shriai visited China to suggest collaboration with China in this project. Because this drug is in the same family of compounds as artemether (which had been approved for production by China) and, therefore, to avoid competition with artemether in the international market, China agreed with the proposal for collaboration in developing the ether derivative.

On October 16, 1987, Director Dr. Godal of WHO/TRD and two other people visited China and discussed this collaboration with the State Administration of Medicine. An agreement was reached, but later, again for unknown reasons, WHO/TDR did not approve the agreement.

Only two years later, China discovered that WHO/TDR had signed an agreement with ACF Company in Holland to develop the ether derivative. The projected budget (1990–1996) was about US$6 million. This included research and development of an aqueous preparation of artelinate (a benzoic acid ether of dihydro-artemisinin), the resources of *Qinghao*, and production of artemisinin. In the course of their research, they found that arteether was highly neurotoxic in animals. They therefore extended the neurotoxicity studies to the Chinese products of artemether, artesunate, and artemisinin. Their studies demonstrated that the neurotoxicity of artesunate and artemisinin was low, and that oil soluble artemether had some neurotoxicity but less than that of the ether derivative. Yet the neurotoxicity in artemether (especially ototoxicity) that they claimed to have found had never been observed in many years of large-scale clinical use.

When a clinical trial on arteether was completed, the total dose for a treatment course used was 960 mg, double the dosage used for artemether. The results of this clinical trial confirmed the positive results found in animal studies by the Shanghai Institutes of Materia Medica carried out years previously. In 2000 arteether was on the market but was not included in WHO's list of essential drugs. The reason for this was unknown, but might have been that it is of the same family of drugs as artemether and artesunate, which were already on WHO's list of basic drugs in 1995, or that its neurotoxicity was higher than artemether and it costs more because of the higher dose used. Other research projects supposedly underway, such as an artelinate water-soluble preparation and artemisinin resource development, had no follow-up reports.

A few years later, although Kunming Pharmaceutical Factory and Guilin Pharmaceutical Factory in China had produced artemisinin, it was difficult to sell the drug on the international market because Chinese production conditions had not reached the international GMP standards. Some large pharmaceutical companies in the Western world collaborated with Chinese pharmaceutical factories that produced *qinghaosu* (artemisinin), by buying their half-finished or completed products at a low price. They then prepared the products for packaging under their own brand name and sold the drug at an expensive price. Certain Western pharmaceutical companies even bought artemisinin material from China and created their own imitation products to sell at a high price. Even today, although the Guilin and Kunming pharmaceutical factories that produce artesunate and artemether respectively have already complied with GMP standards, they are used only for supplying material to certain Western pharmaceutical companies. The world believes that these drugs are the products of only these big companies. In the list of *Qinghao* products that WHO purchased, the names of the Chinese pharmaceutical factories that produced artemisinin were never mentioned. Advantage has been taken of the Chinese artemisinin production industry. Chinese scientists, technologists, other workers, and managerial personnel, who contributed to the discovery and development of *qinghaosu*, sincerely regret that due acknowledgment and credit have not been accorded to them.

Of course, China benefited a great deal during the collaboration with WHO. One benefit was to realize our shortcomings in new drug development and to identify in which direction we should put more effort. WHO also introduced us to new management concepts and models and to new technologies. Selected Chinese *qinghaosu* researchers were sent abroad for further training. WHO also sent experts to China to establish training courses on techniques for measuring drug levels in blood, which was very helpful in studies on the clinical pharmacology of artemisinin.

This collaboration with WHO left the Chinese scientific and technical workers with many concerns, especially those of inadequacy. We had to ask ourselves why we could discover and develop *qinghaosu* and its derivatives, which was a world-class achievement, yet we could not produce a final drug product that met world-class standards from our manufacturing capabilities? China's isolation, not lack of ability or intelligence, was at fault, because we had internal technical protections of our own discoveries and inventions for 7–8 years. Our mistake was not to realize that publication of our data made that information public property, and it was lost to our control and claims of ownership. Cooperation and collaboration might be discussed, but ultimately outside parties could do as they wished with our discovery.

We could not change the past, but we could be more careful in the future. Through this collaboration with the outside world, we learned many useful things, especially to recognize our weak points and to strengthen them. To enter the world market of drug sales and distribution, we would have to comply with the necessary guidelines and regulations such as GMP, GLP, and international patent laws.

CHAPTER 8

Malaria Patients Benefit Worldwide

A. Lessons from the Past: Be Self-Sufficient

Although the collaboration with WHO and Walter Reed in artesunate research was unsuccessful, the consequences were not totally negative, because in the early phase of negotiations, especially when the pharmaceutical factories failed GMP inspection, the *Qinghaosu* Directional Committee had predicted the complexity and unpredictability of future collaborations, if any. The committee decided to stand on its own feet and to rely only on national collaboration within China, as well as thoroughly absorbing and implementing the international standards and techniques suggested by WHO. With financial support of the Ministry of Finance, the research and development of artesunate and artemether were finally completed, and the submitted data successfully passed the evaluation process and obtained approval by the State Drug Administration Department. This was the first time China had developed a new drug according to international standards. This was also a measure of China's new technical strength in drug development.

China's administrative and regulatory system for evaluation and approval of new drugs in the early 1980s was insufficient. At that time, China had no professional clinical pharmacology institutions and few scientific and technical staff trained in clinical pharmacology or preclinical toxicology studies, while knowledge of specific toxicity studies was nonexistent, which meant that the *qinghaosu* (artemisinin) family of antimalarial drugs discovered, developed, and manufactured by China could not be recognized internationally.

The national Project 523 leading group and later the *Qinghaosu* Directional Committee paid serious attention to the safety aspects of artemisinin (*qinghaosu*).[47] In 1981 China had not yet implemented GLP administrative standards and had no laboratory that met GLP standards. But China already understood that for drug toxicity results to be accepted internationally, the experiments had to be performed in laboratories complying with GLP standards. China therefore followed GLP requirements and carried out a *qinghaosu* oil emulsion fourteen-day toxicology study in large animals (monkeys). The result showed that at a dosage three times the clinical therapeutic dosage, there were no toxic effects in monkeys. At six times the clinical therapeutic dosage (48MKD), there was a transient decrease in reticulocyte counts. Only when greater than twelve times the clinical therapeutic dosage (96MKD) did apparent toxic side effects occur. Toxic effects included severe suppression of hematopoiesis, damage to the myocardium, appearance of neurotoxicity, and one-sixth of animals died.[47] The fourteen-day toxicology study with the same dosage groups was repeated in dogs—the results were similar but less severe, and there were no animal deaths. It was concluded that artemisinin was a safe drug for treating drug-resistant falciparum malaria and for severe malaria.

Later in the early 1980s, the *Qinghaosu* Directional Committee, following a research design and protocol suggested by WHO SWG-CHEMAL and with the guidance of visiting experts, carried out research on the artemisinin family of antimalarial drugs according to international standards. This started a precedent of researching and developing new drugs by following international standards and led to a correction of earlier errors.

Professor Li Zelin of the Department of Pharmacology of Beijing Institute of Traditional Chinese Materia Medica examined possible mutation and reproductive adverse effects of the artemisinin type of antimalarial drugs, and found none. The results eliminated any fear that these drugs might cause "development arrest" similar to the thalidomide tragedy in the 1960s. In 1983 Professor Zeng Meiyi described the mechanism of the chemical reaction of artemisinin using ultraviolet spectrophotometric methodology. In 1986 Zhao Shishan established the first high-performance liquid chromatography (HPLC) analysis of artemisinin in China to measure levels of artemisinin in blood and saliva in animals and humans, as well as the *qinghaosu* content in plants. Professor Zhou Zhongming did the first *in vivo* pharmacokinetics study in Chinese patients using artemisinin suppositories. In 1986 Professor Zeng Meiyi developed a high-performance liquid chromatography (HPLC) method to measure lumefantrine, which later was used for blood level determinations.

Professor Teng Xihe of the Department of Pharmacology & Toxicology in the Institute of Microbiology and Epidemiology at the Academy of Military Medical Science headed the long-term toxicology study of the artemisinin family of antimalarial drugs. In order to comply with GLP requirements and while training the scientific and technical staff for the task, he wrote a standardized operation procedure (SOP) for various toxicology experiments and confirmation tests. This systematic approach provided complete, reliable, and accurate scientific data for the research summary report, and won him high compliments from the drug evaluation departments. This was also the first preclinical drug safety evaluation in China that followed and complied with GLP standards.

Professor Song Shuyuang of the Medical Radiology Institute at the Military Academy of Medical Science carried out toxicology studies in animals and found that a large dose of artemisinin lowered the early reticulocyte counts. This observation later became one of the important clinical indicators of toxic side effects in patients using the artemisinin type of drugs.

Professors Song Zhenyu and Yang Shude from the Pharmacology Institute at the Chinese Academy of Medical Science carried out clinical pharmacokinetic studies of the artemisinin family of antimalarial drugs. They established the radio-immunoassay (RIA) and high-performance liquid chromatography with reductive electrochemical detector (HPLC-Ee) methodologies. These measurements served as the scientific basis for designing clinical drug-dosage administration protocols.

Professor Li Guoqiao of the Department of Malaria Research at Guangzhou College of Traditional Chinese Medicine was responsible for phase I clinical studies of the *qinghaosu* type of antimalarial drugs. He tested healthy volunteers to observe side effects and drug tolerance. Phase I clinical studies were unprecedented in clinical research in China prior to this time. To comply with international standards of ethics for clinical trials, the Department of Drug Administration of the National Ministry of Health approved Guangzhou College of Chinese Traditional Medicine to carry out phase I clinical studies. Zhu Yen from the Drug Control Institute at the National Ministry of Health acted as a monitor for the phase I drug testing. The study design and protocol were the same as WHO's protocol for artesunate and artemether phase I clinical studies. This was the first double-blind drug tolerance and pharmacokinetic study adhering to WHO's protocol in China.

To further understand the requirements for international drug registration, Professor Shen Jiaxiang of the State Administration of Medicine instructed the *Qinghaosu* Directional Committee to ask Teng Xihe, Deng Rongxian, Zhao Xiuwen, and Professor Zeng Meiyi to gather and translate the material and data on artesunate and artemether and to present it according to the format and requirements for international new drug registration. The Committee also asked experts and consultants of WHO SWG-CHEMAL for their help in reviewing their data. WHO's experts and consultants made many helpful comments and constructive suggestions.

After more than ten years of collaborative research by related units, the National Ministry of Health in 1987 approved the artemisinin derivatives, artemether oil injection forms, and artesunate aqueous injection forms for production. On September 11, 1987, the National Ministry of Health had a press conference announcing the success of the artemether and artesunate injection forms and artemisinin suppositories, as special drugs for treating drug- resistant falciparum malaria and cerebral malaria; they also announced the commencement of their marketing in China.[48, 49] In 1992 dihydro-artemisinin tablets and combination tablets of artemether with lumefantrine were approved for production in China. From 1998 to 2004, the State Food and Drug Administration issued new drug certificates and new drug production approvals for two new combinations: artemisinin with naphthoquine and dihydro-artemisinin with piperaquine.

From 1985, when China started the improved requirements for evaluating new drugs, to 1995, China has approved fourteen kinds of new drugs, among them seven artemisinin types of new drugs. This demonstrates the positive effects international contacts had on our drug research and development program in China, and it also demonstrates the position of artemisinin types of drugs in China's history of new drug development.

***

B. First Large-Scale Factory Built for Production of Qinghaosu (Artemisinin)

Early in 1978, the National Ministry of Health, the Ministry of Chemical Industries, and the General Logistics Department of the People's Liberation Army jointly ordered the urgent production of artemisinin. Although there were factories already producing artemisinin, two problems needed solving. First, information on availability of *Qinghao* resources was incomplete, and second, the existing facilities were limited in their production capacity. Urgent on the agenda was the need for an industrial-scale facility to manufacture large amounts of artemisinin.

The nationwide *Qinghao* resources investigation in the 1970s had already identified the Youyang District in Sichuan Province as the best *Qinghao* resource. Wild *Qinghao* grew here in abundance, was distributed in dense tracts, was of good quality, and had high artemisinin content. Most of the artemisinin required for scientific laboratory and clinical research at that time was extracted from *Qinghao* herbal plants from Youyang District. Realizing the significance for commercialization of artemisinin and its derivatives, the *Qinghaosu* Directional Committee considered developing and protecting the quality *Qinghao* resources in Youyang District. From 1983 to 1984, the committee organized the financing and sending of scientific and technological workers to Youyang District, to reinvestigate *Qinghao* resources and to look into the feasibility of building a factory in Youyang for industrial production of artemisinin.

The *Qinghaosu* Directional Committee in July 1986 recommended that Sichuan Province begin planning to build an artemisinin production factory in Youyang District. The Sichuan Drug Administration and Provincial Health Department approved the building of Wuling Shan Pharmaceutical Factory. This would be the first artemisinin industrial production factory, designed by China, with an annual production capacity in the amount of tons. Because Youyang was a poor mountainous district with a lack of technical and financial support, to ensure the capacity of the production process of Wuling Shan Pharmaceutical Factory, the *Qinghaosu* Directional Committee coordinated collaboration with Kunming Pharmaceutical Factory. An agreement was signed between them to jointly produce artemisinin. In the signed agreement, Kunming Pharmaceutical Factory was responsible for the quality of artemisinin production and invested 200,000 yuen into Wuling Shan Factory. The *Qinghaosu* Directional Committee also paid the Shandong Institute of Chinese

Traditional Medicine to assist Sichuan Wuling Shan Factory with the manufacturing procedures and equipment necessary for artemisinin production. This was an important contribution to the industrialization of artemisinin production. In July 1987, Wuling Shan Pharmaceutical Factory and the Shandong Institute of Chinese Traditional Medicine signed a contract to "Transfer the Production Processing Technique for *Qinghao* Raw Material." In December 1987, Sichuan Pei Ling District issued a permit for Wuling Shan Pharmaceutical Factory to build its production facility. On New Year's Day in 1988, the ground was broken and construction began. In 1990, because of the involvement and support of leading departments, the building of this large-scale factory for the industrial-level production of artemisinin was completed and production started. This is the first such factory in China and in the world.

The investment in artemisinin production at Wuling Shan Pharmaceutical Factory created favorable conditions for promoting artemisinin and its derivatives throughout the world, and it had important historic significance in the development of artemisinin and its derivatives.

***

C. Science, Industry, Commerce, and International Markets

Research and development of artemisinin and its derivatives was a high-technology achievement, and for it to be accepted internationally required an effort involving science, industry, and commerce. It had initially been under the unified guidance and coordination of the national Project 523 head office and later under the *Qinghaosu* Directional Committee, coordinating with various related ministries and departments.

To develop an international market for the artemisinin family of new drugs was difficult, because of the large investment required, the high international standards, and high technological demands. It was all very alien and difficult for the Chinese pharmaceutical industry and scientific academic research units at that time. Of most importance and concern was the outside world's interest in artemisinin research and development, which was a challenge to the Chinese artemisinin production industry. Within China there was unregulated competition to obtain *Qinghao* raw material and to produce and export the finished product. At this time, and to make matters worse, the *Qinghaosu* Directional Committee, which was established by the National Ministry of Health and the State General Administration of Drugs, was closed down. This occurred on June 13, 1988, at the time when the Chinese registered artemether, artesunate, and artemisinin suppositories were about to enter into the market. From then on, individual departments and units managed research and development of the artemisinin family of antimalarial drugs, and the chain of cooperative effort between science, industry, and commerce was broken.

After the dissolution of the *Qinghaosu* Directional Committee, it was fairly certain that chaotic competition would be the result unless timely measures were taken to reorganize. China's artemisinin research and the production industry would be put in an unfavorable situation internationally, and China would incur significant financial losses.

In November 1987, Deputy Director Guo Shuyan of the National Commission of Science and Technology and Professor Guo Dexi of the State General Administration of Drugs visited Africa. Wang Yushan, Chinese Ambassador to Nigeria, told the visitors of the very positive Nigerian reaction to the news regarding the excellent results of the new Chinese drug artemether in treating malaria. Ambassador Wang suggested to Deputy Director Guo Shuyan that this would be a most opportune time to introduce artemisinin into Africa. He stated that "Africa needs *qinghaosu*, and vice versa, *qinghaosu* needs Africa!" People in Nigeria and other African countries considered *qinghaosu* "an Oriental magical drug and a life-saving drug from China." For Africans artemisinin drugs had become a valuable product to be distributed among friends and relatives.

On returning to China, Deputy Director Guo Shuyan quickly dealt with the matter. He assigned the responsibility for organizing and coordinating a task force to Director Cong Zhong and Chen Chuanhong of the Social Development and Technology Department. Zhou Keding, secretariat director of the earlier national Project 523 head office and later *Qinghaosu* Directional Committee, was the consultant for the task force. In January 1988, the National Commission of Science and Technology sent the General Office of the State Council a report concerning the Chinese production of the artemisinin family of antimalarial drugs. The report also dealt with their sales in the international markets, mentioned several measures to be taken, and assigned the task to the various ministries and commissions concerned. On May 24, 1989, Deputy Director Guo Shuyan called a meeting to discuss the production, sales, and the developing of an international market for artemisinin and derivatives. Participants at the meeting included senior officials and representatives from the National Ministry of Health, the State General Administration of Drugs, Ministry of Economy and Trade, and several pharmaceutical factories (Kunming, Guilin No.1, Guilin No.2, Guangzhou Baiyun Shan).

Deputy Director Guo Shuyan stated that "In order to promote this task, the National Commission of Science and Technology and various ministries and departments concerned have had three meetings and jointly issued documents. These have led to the building of good relationships and the delineation and coordination of the work. Coordination is essential because the outside world is pressing ahead with their own development plans. China still only has the potential for economic benefits, but no actual benefits so far. Several pharmaceutical factories are very concerned because of the inability to export their products leading to the current shortage of capital, and waiting for the opportunity to compete on an equal basis is difficult. I invite all of you to discuss the means for coordination and collaboration within China, and to unify your efforts to deal with the outside world so as to convert the potential benefits into real benefits."[50]

After the meeting, the National Commission of Science and Technology wrote a plan for "Expediting the Promotion, Exportation and Foreign Exchange Earnings of *Qinghaosu* Based Antimalarial Drugs." According to the plan, the National Commission of Science and Technology brought together the three branches of science, industry, and trade to work together as a single body. The aim was to look for international technology collaboration, international drug registration, and trade negotiation; to choose export agents through competitive bidding; and to provide certain investment opportunities as the basis for future profit sharing of foreign exchange earnings.

The plan also requested that there should be a sharing of responsibility among ministries and departments. The General Administration of Custom Services would watch for the exporting of *Qinghao* resources, such as *Qinghao* seeds and materials. The State General Administration of Drugs would supervise quality of production to reach GMP standards as soon as possible. The Ministry of Economy and Trade would take charge of developing international markets. The Ministry of Health would be responsible for promoting artemisinin-based antimalarial products through advertising and by sending malaria treatment and prevention teams abroad to help our allies.

On July 16, 1988, the National Commission of Science and Technology, the State General Administration of Drugs, the Ministry of International Economy and Trade, the Ministry of Agriculture, and the National Ministry of Health jointly issued the "Notice for Expediting the Promotion, Exportation and Foreign Exchange Earning of *Qinghaosu*-Based Antimalarial Drugs." The group also had an operational protocol to use the competitive method for choosing a constructive and better-qualified company for artemisinin international collaboration projects. During August 28–30, 1989, the National Commission of Science and Technology met with nineteen trading companies (including the China MEHECO Corporation, its offshoot, China Science Resource [Ke Hua] Technology Trading Corporation, and China National Trust and Trade Company), and three pharmaceutical factories (Kunming, Guilin No.1, and Guilin No.2). The purpose was to discuss artemisinin international collaboration, to sign collaboration contracts (contracts between the National Commission of Science and Technology and trading companies, and contracts between trading companies and Chinese pharmaceutical factories), to discuss matters related to this collaboration, and to divide trade regions based on the countries each company or corporation had agents in.

This unified organization and coordination of the National Commission of Science and Technology prevented a chaotic situation developing in China. Through public bidding, nine competent trading companies with the rights to export represented the production pharmaceutical factories in China, and could start drug registration outside China, and become involved in clinical studies and technical collaborations. Among these trading companies, CITIC Technology Company and China Science Resource (Ke Hua) Technology Trading Company had the best results in seeking international collaboration. Because of the cooperation between pharmaceutical factories, scientific research units, and commercial trading companies, there was a clear division of responsibility, rights, and profits, and Chinese industry became more active in international competition. After four years of hard work, some of the artemisinin-type antimalarial drugs were registered in twenty or more countries in Southeast Asia, Africa, and Latin America. Drugs were exported to countries or areas that had an urgent need to treat malaria. The international scientific and technology industry collaboration was progressing well. Various forms of scientific and technology collaboration, and commerce and trade collaboration were established with many multinational companies, such as Novartis in Switzerland,

Sanofi in France, and Cadilia Pharmaceutical Limited in India. Developing the international markets for the artemisinin family of antimalarial drugs had made major advances, and China had gained much valuable experience relating to the introduction of new drugs internationally.

With the tendency towards global economic integration, many multinational companies and corporations were merging. For now and in the future, China's competitors were not only within China, but also at international level. China was facing challenges on all fronts, from the use of resources to product development, marketing, and scientific research. Because we had a closed system over many years, we had to learn some painful lessons in commercial and in scientific and technological collaboration negotiations. Meanwhile, we realized that we were inexperienced in developing an international market, and that investment in Chinese companies and corporations was deficient, leading to financial weaknesses. There was a lack of a long-term vision with the emphasis aimed mostly at immediate profits. There was still a very long time needed for us to build our own international sales network. To survive and to expand under the new conditions of the market economy, Chinese commerce and trade corporations and research units involved in artemisinin production needed to continually explore ways to develop an international market.

***

D. China Leads in Developing Qinghaosu (Artemisinin) Combinations

The synthesis of artemisinin derivatives was the first major advance following the initial discovery of *qinghaosu*, and it demonstrated the ability of the Chinese scientific community.

Both artemisinin and its derivatives had a very rapid effect in controlling the clinical symptoms and signs of malaria and in destroying the malaria parasites. After the first dose of the drug, fever and clinical symptoms were reduced or disappeared, and parasitemia also cleared rapidly, but they were not totally eradicated. Treatment had to be continued for seven days, otherwise parasites recrudesced within ten to fifteen days, and symptoms returned. With the majority of patients being from the poor malaria endemic areas, it was very difficult to convince patients to continue treatment for an additional seven days after they were asymptomatic. Also a longer treatment course meant a higher cost. Additionally, some malaria experts felt that, although currently there were no drugs better than the artemisinin family in treating drug-resistant falciparum malaria, resistance might develop if the drugs were widely and repeatedly used. There was concern that another antimalarial would not be discovered to take the place of artemisinin if resistance developed.

Facing this possibility, Chinese scientists in the field of antimalarial drug research again demonstrated their initiative and insight. This led to the second major advance, which was to combine artemisinin with another effective antimalarial drug. Such a combination product could potentiate the effect of artemisinin and possibly prevent recrudescence.

Early in 1967, scientists carried out Project 523 with a clearly defined aim: to find a new drug to treat drug-resistant falciparum malaria. After extensive research and many experiments, they had accumulated practical experience in creating drug combinations. This was exemplified in the development of Malaria Prevention Tablet MP No.1 (a combination of dapsone and pyrimethamine), MP No.2 (a combination of sulfadoxine and pyrimethamine), and MP No.3 (a combination of piperaquine phosphate and sulfadoxine), thus helping to solve the urgent needs of our Vietnamese ally in combat. These were all combinations of drugs that had improved efficacy in preventing and treating malaria in drug-resistant falciparum malaria endemic areas. Compared to the drugs used by the US army at that time, the drugs and treatment protocols provided by the Chinese scientists were more effective, more advanced, more practical, and easier to use. These researchers had not only solved urgent needs and protected the combat strength of our ally and our own troops, but also over the ensuing ten years discovered a group of new antimalarial drugs, of which artemisinin was an example. These new drugs, including lumefantrine, naphthoquine, and pyronaridine, were the material basis for China continuing to lead the world in antimalarial drug research after the 1980s. The most representative US discovery was mefloquine. Originally WHO and the US military had great expectations for mefloquine, and its early development received encouragement and financial support. Further development, studies, and marketing worldwide were undertaken by Roche Pharmaceuticals of Switzerland. It was widely used for treatment and prophylaxis, but resistance in some areas appeared and there were reports of neuropsychiatric side effects. A combination called Fansimef (Fansidar + mefloquine) was also developed and had limited success.

The Institute of Microbiology and Epidemiology of the Military Academy of Medical Science of the People's Liberation Army was the first unit for malaria prevention and treatment research in China. Since 1951 this institute had accumulated experience in preventing and treating malaria in the military forces in Hainan and at the border of Yunnan Province. It had a trained scientific research team of high quality and technological standards. This team was one of the principal research teams involved in antimalarial drug research for Project 523.

Drug chemist Deng Rongxian, now deceased, and researcher Zhang Xiuping from the Shanghai Institute of the Pharmaceutical Industry were the leader and deputy leader respectively of the Project 523 synthetic drugs specialty group. Besides organizing and coordinating the work for members in the group, they each made outstanding contributions based on their own research.

Under the leadership of Professor Zhang Xiuping, the Project 523 synthetic drug team of the Shanghai Institute of the Pharmaceutical Industry, collaborating with Shanghai

Pharmaceutical Factory No.2, synthesized sulfadoxine. And collaborating with Xu Deyu, Shen Nianci and others from the Second Military Medical University created the long-acting antimalarial drug piperaquine phosphate. These two drugs were the main components for the later combinations of Prevention Tablets MP No.2 and No.3 respectively.

Since the start of Project 523, the antimalarial synthetic drugs research unit, headed by Professor Deng Rongxian, had synthesized a large number of new chemical compounds, from which lumefantrine was selected. Researchers Li Fulin and others of the same unit synthesized naphthoquine phosphate. Researchers Teng Xihe, Zhou Yiqing, Jiao Xiuqing, and Wang Yunlin completed the pharmacology and pharmaceutical and clinical studies on these two new antimalarial compounds. Both compounds were approved as new drugs, the former receiving first prize in the National Award for Inventions, while the latter received second prize. Artemisinin and these two drugs received the highest awards ever in China for their discovery. These two new compounds and artemisinin were the material basis for China's antimalarial combination preparations later, assuring China's leading role in the field.

In 1990 the Institute of Microbiology and Epidemiology of the Military Academy of Medical Science created the first artemether-lumefantrine combination, and it was approved for registration as a new drug. In the 1990s, the new combination of artemether and naphthoquine phosphate was registered. The former combination artemether-lumefantrine was jointly developed with Switzerland's Novartis Company (previously Ciba-Geigy). Novartis invested a large amount of money by repeating all experiments according to international standards and the requirements of new drug research. This combination entered WHO's "Essential Medicine List" in 2000. This signified the advanced level of China's research on combinations with artemisinin after the earlier discoveries of *qinghaosu* and then its derivatives.

In 1981 in the Beijing conference on *qinghaosu* and its derivatives, the concern of the possible development of artemisinin drug resistance, after its wide use, was raised and discussed. With clear foresight, the researchers of the Institute of Microbiology and Epidemiology started experimenting for different ways to prevent or delay drug resistance to the *qinghaosu* series of drugs, and other new antimalarials, or to increase the effectiveness of drugs to which resistance had developed.

Researchers Zhou Yiqing and Teng Xihe headed the research team for this project. Based on the earlier urgently developed combinations for assistance to our allies, they started experiments in the 1980s combining artemisinin and its derivatives with other antimalarial drugs. When their proposed cooperation with WHO was turned down with comments such as "you don't have the facilities and ability to perform combination-drug research," they established a new animal model for experiments in 1982 with financial support from the *Qinghaosu* Directional Committee. Using this animal model, various combinations of artemisinin and its derivatives with sulfadoxine-pyrimethamine—lumefantrine, for example—were tested. The combination artemether-lumefantrine was selected as the main research target.[51]

As mentioned before, artemether was the result of cooperative efforts between the Shanghai Institute of Materia Medica of the Chinese Academy of Sciences and Kunming Pharmaceutical Factory. It passed the drug evaluation committee in 1981 and was a new drug developed solely by China. So was lumefantrine. Lumefantrine was an antimalarial drug with a new chemical structure and no cross-drug resistance with chloroquine. It obtained its Chinese "New Drug Certification" in 1989, and was first manufactured by Kunming Pharmaceutical Factory and then by Zhejiang Xinchang Pharmaceutical Factory. Lumefantrine was a slow-acting drug for treating malaria, but in combination with the fast-acting drug artemether they complemented each other, and therefore the combination increased the therapeutic effectiveness.

The Institute of Microbiology and Epidemiology of the Military Academy of Medical Science carried out all the experiments required for new drugs in China with the combination of artemether-lumefantrine. In 1992 this combination was approved as a "category three new drug,"and obtained its "New Drug Certificate" and "Approval for Production." At this moment, the first artemisinin combination new drug was born in China.

***

E. *International Collaboration in Developing New Drugs*

The problem facing the new antimalarial drug artemether-lumefantrine combination was finding a market for it. In China the market for antimalarials was very small, and malaria endemic areas outside China where this drug was needed were generally very poor, and therefore sales would not be profitable. International pharmaceutical companies were not interested in this drug. To show the full effects and value of this combination drug and to benefit the poor in malaria endemic areas in the world, the drug needed to enter into the international markets. Before 1990 the leading departments of the Ministry of Science and Technology and the National General Administration of Medicine had tried to export the artemisinin family of antimalarial drugs to bring profits back to China, but to no avail. Now trying to export artemether-lumifantrine to the world was a real and practical problem.

Since 1990 the CITIC Technology Company had collaborated with the parties involved with the artemether-lumifantrine combination, namely the Institute of Microbiology and Epidemiology of the Academy of Military

Medical Science, Kunming Pharmaceutical Corporation, and Xinchang Pharmaceutical Factory of Zhejiang Medicine Co. Ltd. It acted as the sales agent for artemether combination tablets (artemether-lumifantrine combination) in negotiating cooperation with Ciba-Geigy (now Novartis) in Switzerland. It had the support of five Chinese ministries and commissions, including the National Commission of Science and Technology (now Ministry of Science and Technology). After fifteen years of international scientific and technological cooperation and a large investment from Novartis to repeat all the work needed on this combination drug, it was finally registered in seventy-nine countries and regions, and available for sale in twenty-eight countries. This was the first Chinese drug ever to go into the world as the result of research meeting international standards through collaboration with an internationally well-known pharmaceutical company. As for our artemisinin and its derivatives research, this artemisinin combination again signified our ability to do research at the international level. At the same time, it also represented a successful model for getting a totally Chinese product, under standardized manufacturing management, from raw material to end product and into the world through Western accepted means, and gaining financial benefits for us. The combination artemether-lumifantrine product, under the Novartis Pharmaceutical Company trade name "Coartem," gained a high reputation internationally. It was entered into WHO's "Essential Medicine List" in 2000.

It was worthwhile to review and summarize the process of international cooperation on artemether-lumifantrine. Its success held important clues for later generations. It showed the Chinese new drug research units, the production industry, and the commerce industry how to develop international cooperation or collaboration to create a new pathway for Chinese drugs. Each product is different, and one model would not apply to all, but there was one unavoidable imperative: all aspects of drug research, development, registration, and marketing must adhere to international standards.

On March 9, 1990, a confidentiality agreement was signed between China and the Novartis Company of Switzerland. The Chinese side was the single collaborative body for the science units, production industry, and commerce units, headed by the Social Development and Technology Department of the National Commission of Science and Technology. It consisted of the CITIC Technology Company, the Institute of Microbiology and Epidemiology of the Military Academy of Medical Science, Yunnan Kunming Pharmaceutical Factory, and Zhejiang Xinchang Pharmaceutical Factory. On April 29, 1991, both parties signed "Phase I Preliminary Joint Venture Agreement." On May 16, 1991, "Agreement for Joint Application of Patent" was signed. On September 12, 1994, the twenty-year "Cooperation Agreement" was signed and approved by the National Commission of Science and Technology.

According to the "Cooperation Agreement," the artemether-lumifantrine combination needed to be reevaluated following international standards in various parts of the world. Novartis requested both parties to repeat preclinical studies, clinical trials, and a complete review of all research data. The conclusion reached was that data of all our initial experiments and studies coincided with the results of the repeat studies by an international research company. There were two key questions that determined the success of the cooperation.

After signing the "Phase I Preliminary Joint Venture Agreement," Novartis tried to assess the research level and ability of Chinese scientists by asking two questions. This in reality was an examination. The first question related to the accuracy and reliability of methods for measuring blood levels of the combination drugs, and the second to the credibility of the therapeutic efficacy of the combination.

Question one was directly related to the explanation of the clinical therapeutic efficacy of the combination. Neither artemether nor lumifantrine was water-soluble. The *in vivo* pharmacokinetics and the effectiveness of artemether had been confirmed earlier. Novartis could not deny this. But lumifantrine was a new antimalarial drug and little known outside China. In addition, Novartis found that the testing equipment and instruments in our laboratory were an assemblage of pieces from four different manufacturers. It was understandable that Novartis should be so cautious. Novartis also paid a well-equipped laboratory outside China to establish a method for blood drug-level determinations. Using the same blood samples of volunteers, they did a parallel comparison with our method of blood level testing. The Chinese side strictly followed international requirements, accumulating a large amount of data to check the accuracy, the precision, and the reliability of the methods we had established, and supplied the quality control range of our testing method. But because our test results were fourteen times higher than their results, Novartis sent their personnel to our laboratory to carry out an on-site inspection, to check the managerial organization and the technical operational manuals. Novartis also sent us unknown samples to run parallel comparative tests with their laboratory. The test results from both laboratories were identical, finally demonstrating that our experimental data were accurate and reliable, and that the results of the laboratory contracted by Novartis to rerun the volunteer blood samples were erroneous. Consequently Novartis had to admit that the testing method we had developed was up to international standards.

Meanwhile, to answer the second question, both parties again organized clinical teams at Hainan Island to repeat clinical trials with the new combination. The results were identical to our initial trial results.

After these two tests, Novartis changed its attitude regarding the ability and quality of the Chinese scientists. The artemether-lumefantrine combination joint venture advanced from the "Phase I Preliminary Joint Venture Phase" to the "International

Collaboration Phase," which was officially signed by both parties. In this process of academic and technological comparative assessment, Professors Teng Xihe, Zeng Meiyi, Zhou Yiqing, and Lu Jiliang made significant contributions. Their outstanding work not only promoted international collaboration but also regained the dignity and honor of Chinese scientists. The standard of Chinese technological-scientific researchers in new drug development and their ability to conceive the artemether-lumefantrine combination gained them international respect.

During the fifteen years of the international joint venture with the artemether-lumefantrine combination, which had been initiated under the Chinese regulatory requirements for new drug development, the combination successfully fulfilled WHO's GMP (Good Manufacturing Practice), GLP (Good Laboratory Practice), GCP (Good Clinical Practice) standards, and the EU (European Union) regulatory standards for new drugs. To ensure global use of this drug, an application was made to the US FDA for approval.

An application for a drug patent for this combination has been submitted to sixty-four countries and districts, has received patent protection in forty-nine countries, including the EU countries, the US, and Canada. This was not only the first Chinese-developed drug to obtain an international patent, but also the first artemisinin combination to have received international patent protection in more countries than other patented artemisinin combinations. It is also one of only three patented drugs on WHO's "Essential Medicine List" in the last twenty-five years. The success in applying for international patent protection has built a solid foundation for international collaboration and for international competition with other well-known trade-named drugs. After being used to treat more than five thousand patients in international, multicenter clinical trials, having its quality control perfected for production, and being tested in evidence-based studies, this artemether-lumefantrine combination is now chosen as the first-line antimalarial drug by many African countries. And it is recommended as the drug for treating malaria by WHO, Doctors Without Borders (MSF), and the Global Fund to Fight AIDS, Tuberculosis and Malaria (GFATM).

It is worthwhile to mention that the international collaboration on the artemether-lumefantrine combination was the correct decision, considering the amount of money needed for entry into the international markets, application for international patent rights, and for international new drug registration. With no short-term financial gain, companies in China would not have the courage and endurance to proceed with this. Without international collaboration, this research product would have been put aside and the benefit to malaria patients worldwide would have been lost.

***

F. Another Accolade for Qinghaosu (Artemisinin)

The research group of Guangzhou College of Traditional Chinese Medicine, headed by Li Guoqiao, was another major force in demonstrating to the world outside China *qinghaosu*'s effectiveness in treating malaria. In the forty years since they accepted collaboration in Project 523 in 1967, the group has continually carried out antimalarial drug research. In 1974 they accepted responsibility for clinical trials with *huanghaosu*, and for the past thirty years have carried out clinical research on *qinghaosu* (artemisinin), the derivatives and combination drugs.

In 1974 in collaboration with the Yunnan Institute of Materia Medica, they made the first observation that *huanghaosu* (*qinghaosu*) was highly effective, with a rapid onset of action, and had low toxicity but a high short-term recrudescent rate.

After the Chengdu meeting and with the arrangement of the national Project 523 head office, they became the leading collaboration unit for *qinghaosu* clinical research. Later they were responsible for the clinical trials on *qinghaosu* derivatives artemether, artesunate, and dihydro-artemisinin, and artemisinin (*qinghaosu*) suppositories. The results of their success in treating cerebral malaria with *qinghaosu* were the basis for WHO's decision to make artesunate the first-line drug in treating cerebral malaria. In many clinical trials from 1984 to 1988, with more than one thousand malaria patients, Li Guoqiao compared a three-day, five-day, and seven-day treatment course, to conclude that the seven-day course could increase the cure rate to 95 percent. This research in prolonging the artemisinin treatment course changed the perception that artemisinin always had a high recurrence rate. Their suggested seven-day course protocol was recognized by WHO as the standard treatment course for the *qinghaosu* (artemisinin) family of drugs in treating falciparum malaria.

In the 1980s, with artemisinin types of antimalarial drugs now available as commercial products, the number of pregnant patients with severe malaria using artemisinin types of drugs increased. On the one hand, because of its rapid action, artemisinin decreased tremendously the rate of intrauterine fetal demise and spontaneous abortion, and effectively saved the lives of mothers and fetuses. But on the other hand, did artemisinin affect the development of the babies born to these women? To answer this question, they followed the children (age three to six years) born to twenty-three pregnant patients with cerebral malaria treated with artemisinin for many years, and found no morphological abnormalities or mental retardation. Therefore in 1989 the team concluded that artemisinin types of drugs were appropriate to use in treating uncomplicated falciparum and severe malaria in women in the second or third trimester of pregnancy.

In 1990, seeing a study by an Indian researcher on the effects of artemisinin on gametocytes of *Plasmodium vivax*, they studied the effects of artemisinin on the infectivity of *P. falciparum*, with a grant from Roche Asian Research Foundation. From 1991 to 1995, they were invited to Vietnam to help treat patients with cerebral malaria. From bone marrow smears, they studied the effect on the development of early gametocytes. Within three years, they established a method to test the infectivity of *Anopheles* mosquitoes. After dissecting thousands of mosquitoes in this study, they found that artemisinin not only restrained the growth and killed the mature *P. falciparum* gametocytes in mosquitoes, but also rapidly destroyed gametocytes in the earlier stages of development. In the not yet infective gametocytes that had just entered the peripheral blood circulation of patients, artemisinin stopped their maturation and their becoming infective. Regarding restraining and killing the mature gametocytes in the peripheral blood, their infectivity rate was lowered to 70 percent and 100 percent (total loss of ability to infect) on day 7 and day 14 respectively after drug administration. This important discovery corrected the 1978 inaccurate conclusion that *qinghaosu* (artemisinin) had no effect on gametocytes. This false statement had been made in the early clinical trials with artemisinin because gametocytes persisted for a long time in patients' peripheral blood smears in these studies. With this further study into the effect on the infectivity of gametocytes by artemisinin, a new aspect of the antimalarial effect of artemisinin was recognized. This unique gametocyte restraining and killing effect of artemisinin is not seen in any existing antimalarial drugs to date. This discovery suggests an alternative method for controlling the spread of malaria by eliminating the infectivity of gametocytes.

Because the number of malaria patients in China was relatively low, the team established a research base abroad in the early 1990s. Through their clinical trials and research projects at home and abroad, this academic unit promoted artemisinin drugs and helped develop a market for the drugs. With the increased demand for these antimalarials, our artemisinin manufacturing industry also benefited.

It has been thirty years since the discovery of artemisinin and more than twenty years since the commercialization of artesunate and artemether. But these drugs are still not widely used in malaria endemic areas. One of the reasons is the long treatment course and the inconvenience and cost to the patient. One possible way to solve this problem is to consider using a combination of drugs. The ideal antimalarial combination should require a short treatment course, use compounds with a better antimalarial effect and have a low cost. In order to help patients in poor malaria endemic areas in the world to use the Chinese-produced effective, inexpensive, and convenient artemisinin antimalarial drugs, Li's team continued their research with unique new antimalarial combinations.

In the 1980s, Li Guoqiao's research team started clinical trials with various combinations of drugs with artemisinin. They combined artesunate and the long-acting piperaquine phosphate, the two new research products of Project 523, and obtained better treatment results. To understand why this team decided to completely devote their research to new combinations, it is necessary to look at the first time they were invited to Vietnam to treat cerebral malaria with artemisinin.

In 1989 Vietnam was at the stage of economic reconstruction when large numbers of people were moving into areas where malaria was prevalent. *Plasmodium falciparum* in Vietnam was highly drug-resistant, resulting in many cases of severe malaria and a significant death rate. Dr. Keith Arnold, who previously had worked with the US army in antimalarial drug research and was at that time with Roche Asian Research Foundation, was carrying out clinical research with mefloquine in Ho Chi Minh City. He had contacted Li Guoqiao's research team in 1980 and had become a research colleague and good friend. He had also personally visited Li's clinical research base in Hainan Island and had a deep understanding of Li's work. He recommended to Professor Trinh Kim Anh, a malaria expert and director of Cho Ray Hospital in Ho Chi Minh City, that he invite Li Guoqiao to Vietnam to share his expertise in malaria treatment and his extensive experience in the use of artemisinin.

In 1991 Li Guoqiao accepted the invitation and went to Ho Chi Minh City to introduce the use of artemisinin for treating severe malaria. The relationship between China and Vietnam at that time was still not normalized. The Vietnamese minister of health privately visited Li's team at their hotel to discuss malaria control. Li Guoqiao suggested the immediate use of artesunate manufactured by China's Guangxi Guilin Pharmaceutical Factory in all epidemic and endemic areas of Vietnam. Once this was done, the following year the Vietnamese malaria mortality rate decreased tremendously, and the artesunate sales of Guangxi Guilin Pharmaceutical Factory dramatically increased. Two years later, most pharmacies in Vietnam, large and small, had replaced quinine with artesunate.

During this time, Li's team felt that the best way to help Vietnam to treat severe malaria patients was to find an artemisinin combination with a low cost and short treatment course. This could be widely used in urban and rural clinics and hospitals to decrease the incidence of severe malaria and hence decrease mortality rate. In addition, with the gametocyte restraining and killing effects of artemisinin, transmission could be halted, leading to a lowering of the intensity of the epidemics and decreasing the malaria incidence rate. They therefore tried to use the combination of artesunate and long-acting piperaquine phosphate that had previously shown good results in their initial trial in Hainan. To their surprise, a single dose of this combination had only about a 70 percent cure rate when used in Vietnam, while in the Hainan malaria area it had a cure rate of 95 percent. Li realized that the use of large amounts of antimalarial drugs during the Indochina War in the 1960s and 1970s had rendered Vietnam the most severe multi-drug-resistant falciparum malaria endemic area in the region. There was also the possibility of different strains in different areas. Following this experience, Li repeated the clinical trials with various drug combinations in Vietnam, and also repeated his laboratory studies. After five years of intensive work, the first dihydro-artemisinin-piperaquine phosphate combination and a new antimalarial drug, "Malaria Tablet CV8," was registered and approved for production in Vietnam in 1997. Two years later, it became the first-line antimalarial drug in Vietnam. The name "CV8" signified the China-Vietnam collaboration, with C for China, V for Vietnam, and 8 for the eighth of ten drug combination protocols tested in Vietnam. At the same time, they had tried a new combination of dihydro-artemisinin and naphthoquine phosphate that needed to be taken twice in one day. After completing the preclinical studies, the new combination was approved for clinical trials by the State Food and Drug Administration. These two new combination drugs (dihydro-artemisinin and piperaquine phosphate; dihydro-artemisinin and naphthoquine phosphate) and the later combination dihydro-artemisinin and piperaquine all obtained the Chinese "Invention Patent Certificate."

During the registration and manufacture of "Malaria Tablets CV8" in Vietnam, Dr. Allan Schapira, the WHO representative in Vietnam, Cambodia, and Laos, followed this matter closely. In 1998 he visited Guangzhou College of Traditional Chinese Medicine's Department of Tropical Diseases, to review in detail the research on the development of CV8 and its clinical trial results.

In early 2000, WHO/TDR invited Li Guoqiao to attend the Prevention and Treatment of Drug- Resistant Falciparum Malaria Conference in Thailand as a temporary consultant. Li was asked to present research on CV8 malaria tablets. In the same year, WHO/TDR requested a secrecy agreement be signed with Li Guoquiao to hand over all research data on CV8 to WHO/TDR for evaluation. In 2001 WHO/TDR suggested to Li some ideas for improving the CV8 combination. In May 2001, WHO's Dr. Allan Schapira and Dr. Jeremy Farrar of Oxford University Clinical Research Unit in Vietnam went to Guangzhou to discuss in detail the improvements suggested. At this time, Li Guoqiao already had the newly improved CV8 samples ready for them.

They were surprised to find that the components and their relative ratio in the improved CV8 tablets were identical to what they had in mind for discussion! They temporarily named the new combination Artekin, and brought back with them drugs for two hundred patients for clinical trials by the Oxford University research team in Vietnam.

Because the basic cost for making Artekin was relatively low, Dr. Allan Schapira was seriously interested. Although Coartem, by Switzerland's Novartis Company, was already on WHO's "Essential Medicine List" at the time, it was expensive. Dr. Allan Schapira hoped that in the future Artekin could replace chloroquine in public clinics and hospitals in Africa to reduce the high mortality rate and incidence of drug-resistant malaria in that country. Later Dr. Allan Schapira was promoted to malaria consultant and representative for the West Pacific region and later to a principal member of the WHO head office responsible for malaria control. Immediately after heading and organizing the WHO's antimalarial drug development conference in Shanghai in November 2001, he made plans for a WHO meeting in Guangzhou in March 2002 to discuss expediting Artekin to meet international standards. This was yet another meeting organized by WHO SWG-CHEMAL for the antimalarial artemisinin, after their previous two meetings with China in Beijing in November 1981 and May 1985. The organizers of the Guangzhou meeting in April 2002 to discuss Artekin, a dihydro-artemisinin and piperaquine phosphate combination in tablet form, invited twelve international authoritative experts in the field to help Artekin obtain international registration and to be listed in WHO's "Essential Medicine List," and to become the first-line antimalarial drug in the world.

To organize the WHO Guangzhou meeting on Artekin, Dr. Schapira met with Guangzhou City Deputy Mayor Lin Yuanhe and leaders of the State Food and Drug Administration, and the State Administration of Traditional Chinese Medicine. To establish GMP conditions for manufacturing Artekin, Dr. Schapira was accompanied by a GMP expert from the US Pharmacopoeia Committee to inspect Guangdong Medi-World Pharmaceutical Co., Ltd. There were three visits in two years before the GMP expert completed his inspections and stated that the factory was capable of manufacturing Artekin according to the GMP standards. To carry out Artekin research to international standards, Dr. Schapira suggested Li Guoqiao and Professor Nick White collaborate with WHO/TDR to jointly apply for Medicines for Malaria Venture (MMV) grants. They were successful in obtaining a grant of US$3.5 million.

Until the end of 2002, comparative clinical studies of Artekin with other antimalarial drugs (including Coartem, a combination of artemether and lumefantrine) were carried out in malaria endemic areas worldwide. There were five thousand cases treated with Artekin and a cure rate reached of 97 percent. Regarding the therapeutic effectiveness, side effects, cost, and the ease to use, Artekin was considered an ideal to perfect antimalarial drug.

Li Guoqiao was the leader of the artemisinin antimalarial research team for artemisinin drug combination research for more than ten years, continuing the positive spirit of the initial Project 523. In recent years, Li has brought some ex-Project 523 members to Guangzhou to extend and expand his clinical research into other areas of drug development, such as improving the quantity and quality of *Qinghao* herbal plant cultivation, and improving extraction methods and production capacity for artemisinin, its derivatives, and combination products. He also reorganized basic studies in pharmacology and toxicology and established a unit that complied with all the required international standards of GAP (Good Agriculture Practice), GLP (Good Laboratory Practice), GMP (Good Manufacture Practice), and GCP (Good Clinical Practice). Regarding new drug research, development, and production, Li Guoqiao established a completely self-sufficient organization with supplementary specialty teams. These specialty teams comprised researchers from academia as well as workers from the manufacturing sector. The purpose was to discover, develop, register, and market new artemisinin combinations using China's own scientific and technological expertise. In this way, appropriate benefit and credit will be accorded to China as the originator of this whole class of new antimalarials.

Since November 2003, in clinical trials in some highly malaria endemic areas in Cambodia, a new combination (Artequick, artemisinin and piperaquine) has shown great potential for the rapid control and eradication of malaria. After the results of further and larger-scale clinical trials become available, this new combination may suggest an alternative way of thinking about methods to reverse the severity of and quickly bring under control malaria epidemics. This outcome is feasible, and could be accomplished if this new combination drug is used extensively in all public clinics, hospitals, and health care centers in malaria endemic countries. Chloroquine is no longer as useful as it was in the past, and new artemisinin combinations should replace it as drugs of first choice. If this should come about, another chapter in China's contribution to malaria control, and possibly even eradication, would be written, giving further recognition and praise for our discovery of *qinghaosu* (artemisinin).

CHAPTER 9

New Revelations from a Review of the Past

The former leaders, scientists, and other workers involved in Project 523 were not only proud, but also very emotionally moved and uplifted when their achievement in discovering artemisinin (*qinghaosu*) in an herbal plant and developing it to several finished and marketable products was reviewed and publicly acknowledged.

A question frequently raised is, "how, during that turbulent, unstable period in our history, namely the Cultural Revolution, with science and intellectual endeavors repressed, with limited technical resources and equipment, could China have made such important scientific progress?" Within two years, to assist our ally, there was large-scale production of three antimalarial combination tablets—Malaria Prevention Tablets No.1 (dapsone and pyrimethamine), No.2 (sulfadoxine and pyrimethamine), and No.3 (long-acting piperaquine phosphate and sulfadoxine). Within three years, *qinghaosu* was extracted from the Chinese herbal plant *Qinghao* and shown to be fast-acting, very effective, and safe for treating malaria. Within four years, it was being produced in large quantities for use in Cambodia, and within seven years it passed the Chinese new drug evaluation process and was approved for production. The discovery and development of *qinghaosu* (artemisinin) by China have been a major accomplishment in antimalarial drug research in the latter half of the twentieth century.

The following is a brief review of Project 523's research program on *qinghaosu* (artemisinin) to be used as a reference by those interested.

***

A. The Concern of National Leaders

Project 523 was an urgent mission to assist Vietnam in its military conflict with the United States by preventing and treating malaria in its armed forces. Support for the project came from the highest authorities in China, Chairman Mao Zedong and Prime Minister Deng Xiaoping. In 1969 Chairman Mao personally read the progress reports from all participating organizations. National leaders Zhou Enlai and Li Xiannian frequently made important suggestions regarding the project. With the concern of the nation's leaders, Project 523 received attention, guidance, and support from all government levels, from ministries, commissions, provinces, cities, and districts. This was the authority needed to enable the project to be completed successfully. Because of this, even during the special years of the Cultural Revolution, when all other scientific activities came to an almost complete stop, Project 523 was guaranteed adequate manpower, materials, and finance from all units and all departments at all levels.

After the first Project 523 meeting to discuss collaborative activities among units, there were significant changes within the government, which also led to changes in the leadership and other personnel of the project units. Also the initial three-year research plan was coming to an end, and a new plan for the coming years was needed. Prime Minister Zhou Enlai reviewed the three-year progress report and the related problems, which had been submitted jointly by the National Ministry of Health, the Ministry of Fuel and Chemical Industries, the Chinese Academy of Sciences, and the Logistics Department of the People's Liberation Army. In 1971, based on instructions from

Prime Minister Zhou Enlai, the State Council and the State Central Military Commission jointly issued State Document (71) No.29, which readjusted the composition of the Project 523 leading group. The National Ministry of Health was the leader, Logistics Department of the People's Liberation Army was the deputy leader, and the Project 523 head office remained in the Military Academy of Medical Science. In May 1971, the national Project 523 leading group met in Guangzhou City to design the five-year research plan, 1971–1976, to strengthen the leadership and define responsibilities.

In 1970, the Indochina War became more complicated. After the military coup by the Khmer Rouge in Cambodia, King Sihanouk, who was in exile in Beijing, recommended to Premier Zhou Enlai an antimalarial regimen suggested by his personal French physician, Amrish, which was similar to our Malaria Prevention Tablet No.2 (a combination of sulfadoxine and pyrimethamine). On May 28, 1971, Premier Zhou handed this regimen to Xie Hua of the Military Control Commission of the Ministry of Health, and to Dean Wu Jieping of the Chinese Academy of Medical Science. Premier Zhou requested the Chinese Academy of Medical Science and the Military Academy of Medical Science to try this antimalarial drug in falciparum malaria endemic areas of Hainan and along the border of Yunnan Province. He further stressed that "If this works, we can supply large amounts of our own MP No. 2 tablets to the Indochina combat zone, where they are suffering from malaria now." Premier Zhou's suggestion was timely and important and further solidified State Document (71) No.29, which emphasized, to leaders at all levels, Project 523's main mission of helping our ally. This further clarified the direction of Project 523 and greatly encouraged all participants in their work.

The new five-year research plan appropriately adjusted the research targets, and Project 523's research entered a new development phase, with important products appearing one after another, including *qinghaosu*.

On February 25, 1973, the national Project 523 leading group representing its four component members (the National Ministry of Health, the Ministry of Chemical Industries, the Logistics Department of the People's Liberation Army, and the Chinese Academy of Sciences) reported to the State Council and the State Central Military Commission the situation according to State Document (71) No.29, and the progress of the five-year research plan. At the time, there was an epidemic outbreak of malaria in five provinces in southern and central China. The report therefore suggested that the drugs created by Project 523 should be widely used in the epidemic and endemic areas in China to control and prevent malaria, in addition to helping our ally. The report also suggested a national Project 523 meeting to review the progress of the five-year research plan so far, and to identify targets for the remaining three years. Li Xiannian and other national leaders approved the suggestions. In May 1973, the national Project 523 leading group met in Shanghai for the aforementioned purpose.

In 1976 the five-year research plan established in 1971 expired. Although the Indochina War was over, many research projects of Project 523 were not yet completed, but some were very close to getting important results. Facing this situation, in March 1977, the national Project 523 leading group called a nationwide 523 meeting in Beijing to report on the situation, to summarize research progress, and to establish the next four-year research plan for 1977–1980. After the meeting, the leading group reported back to the State Council and the State Central Military Commission regarding this nationwide 523 meeting. The report again stressed that malaria prevention and treatment were a national medical research project and requested the close collaboration of the health, science, chemical industry, and military departments. Vice Chairman Li Xiannian concurred on the importance of the task. The summary of this 523 meeting was sent to all related units in provinces, cities, districts, military districts, and units affiliated with various ministries, commissions, and military departments.

From the beginning to the end, Project 523 had always been rated as a national military project, but it was also listed as "scientific and technical development project No. 321" by the State Planning Commission.

***

B. Collaboration Changed Disadvantage into Advantage

In the 1970s, compared to the West, China's new drug research, especially new antimalarial drug research, was far inferior in terms of its research budget and scientific commitment. At the time, China lacked the required scientific professionals, had out-of-date equipment, and new drug research had never been a main research subject.

In order to meet the urgent need of helping our ally to be combat-proficient, we had to supply effective antimalarial drugs in the shortest possible time. The time saved by a quicker supply of drugs resulted in an increase in the efficiency of the military forces. China needed to organize the fifty to sixty scientific, educational, clinical, and pharmaceutical units available in the country into a large collaborative force involving multiple departments, units, and specialties. The country had to make use of equipment scattered throughout various departments and units and to gather together five hundred to six hundred scientific and technical personnel of related specialties into a united front. There was the need to have a management system to allocate budgets; unify a research plan; divide tasks between appropriate units; coordinate collaboration between departments, units, and specialties; and to ensure a close link between laboratory experiments, clinical studies, and the pharmaceutical production process. The project unified individual departments, units, and personnel into a single structure with all advantages necessary to rapidly supply antimalarial drugs to our ally, as exemplified by Malaria Prevention Tablet No.2 (sulfadoxine and pyrimethamine).

After completion of the laboratory studies by the Military Academy of Medical Science, large amounts of the drug were needed for clinical trials. The component sulfadoxine of the combination was not available in China or in Hong Kong. The Ministry of Chemical Industries (now State General Administration of Medicine) and the national Project 523 head office sent members to Shanghai for assistance. The director of Shanghai Pharmaceutical Factory No.2 decided to stop production of some other drugs, thereby freeing the equipment for producing sulfadoxine. The amount produced was enough to make Malaria Prevention Tablets No.2 in time for expanding the clinical trials. In 1968 the national 523 head office organized simultaneous multicenter clinical trials in Hainan and Yunnan. These were carried out by clinical, hygiene, and research personnel from the Military Academy of Medical Science; Shanghai and Guangdong units related to production and scientific research; the Military Research Institutes of Medical Science affiliated to Guangzhou and Kunming military districts; epidemic prevention brigades and related military medical colleges. This extensive collaboration saved one to two years of valuable research time, and the new drug was ready in time to supply the military forces abroad.

As for the research and development of *qinghaosu* (artemisinin), it was like a relay race. The Beijing Institute of Traditional Chinese Materia Medica had a good start in extracting *qinghaosu* (artemisinin), but then ran into difficulty. The Shandong Institute of Chinese Traditional Medicine and Yunnan Institute of Materia Medica started late, but had a smooth run, picked up, and raced forward like the later runners in the relay team. Without the latter two units' participation making such good progress, *qinghaosu* research might have terminated prematurely or been delayed for at least a few years. As for the Yunnan Institute of Materia Medica, its laboratory studies were successful, but the institute lacked experience in clinical studies and could not enroll enough cases for a clinical trial. At the time the high malaria season was about to finish, a decision was made to have Li Guoqiao's team, which happened to be working in Yunnan, to temporarily take over. If their two-month clinical trial with *huanghaosu* had not confirmed its efficacy, there would not have been enough evidence to convince the Project 523 head office to proceed aggressively with the *qinghaosu* research and its ultimate successful outcome.

The development of artesunate was another classic example, after artemisinin, where all the available manpower cooperated on a successful research project. When WHO suggested first developing artesunate among the artemisinin derivatives, the *Qinghaosu* Directional Committee immediately arranged for collaboration between specialists. Eight units from the Institute of Materia Medica, Chinese Academy of Medical Science, and experts in other specialty fields within China followed WHO's requirements for new drug development. The necessary research was quickly completed, artesunate was approved by the Ministry of Health, and a "New Drug Certificate" was issued to all eight units.

***

C. Combining Widespread Efforts to Achieve Success

It was difficult to prevent and treat drug-resistant falciparum malaria in Southeast Asia. The key was how to solve the drug resistance problem. With the war situation at the time and to protect the combat strength of our troops and that of our ally, China needed to find an effective antimalarial drug in a short time. A quick way to solve this problem was to combine existing antimalarial drugs, but resistance was expected to develop to these drug combinations after three to five years of use.

To possibly really resolve the problem for treating and preventing drug-resistant malaria at its root, it may be necessary to follow a different path by finding new drugs with new chemical structures. This should be the new direction for general research planning and designing, as well as the target-specific projects.

For example, while researching on combinations of existing antimalarial drugs, attention should be given to screening new synthetic compounds with no drug resistance for their antimalarial effects. Use the experience of the outside world in finding new antimalarial synthetic compounds, but do not repeat their path of research. In the ten years of Project 523, not only did we find a few new antimalarial synthetic drugs, but we also had a breakthrough in our traditional Chinese medicine research.

The research for finding a new antimalarial drug from Chinese medicinal herbs operated on the principle of combining information from all sources, scattered near and far, throughout the whole country. It started by focusing on the Chinese herb *Radix dichroae* and its β-dichroine content, of which the antimalarial effectiveness, side effects, and chemical structure were already known. The project was to invest more scientific and technical manpower to try to overcome the side effects of vomiting. After *Qinghao* and *Qinghao* extracts were found to be promising, this project was readjusted, reorganized, and redistributed; and manpower and research efforts went into studying all aspects of *Qinghao*. Following a principle of short-term and long-term goals, an initial effort was made to solve the short-term recrudescent problem and drug-resistant problem to assist our ally. The short-term projects included use of local raw plant material, simple-to-make pharmaceutical preparations, drug dosage adjustment, modification of drug preparation forms, prolongation of treatment course, and treatment protocols with combination of drugs. The long-term projects included establishment of *qinghaosu*'s chemical structure and generation of *qinghaosu* derivatives. The short-term projects led to the creation of *Qinghao* tablets with high efficacy, rapid onset of action, low toxicity, simple production process, and low cost. They could be made with locally available material and in large amounts, which solved the potential drug shortage problem during wartime, and for *en masse* malaria prevention and treatment in the population at large. The long-term projects resulted in establishing artemisinin's chemical structure, the formulations artemether and artesunate with increased solubility, higher therapeutic efficacy, and ease of use. They have become the only drugs recommended by WHO to be widely used in its "Roll Back Malaria Initiative."

***

D. Combination of Traditional and Western Medicine to Arrive at the Same Destination by Different Pathways

In the research plan established by the first Project 523 collaboration meeting in 1967, a decision was made in the search for a new antimalarial drug with no drug resistance, to combine the resources of Chinese traditional medicine and Western medicine. Besides allocating a certain amount of manpower into researching new synthetic antimalarial drugs, more efforts and hope were given to finding new antimalarial drugs from the Chinese traditional medicinal heritage. From the beginning, Project 523 had mobilized more manpower to research Chinese medicinal herbs by reviewing generations of medical records, interviewing residents of malaria endemic areas, and collecting open and secret treatment regimens. Initial screening of seventy thousand regimens resulted in five thousand traditional regimens, from which twenty were selected for laboratory experimentation and clinical studies. The most promising from these twenty were chosen for extensive in-depth exploration by using all scientific and technical expertise and equipment available. *Qinghao* (also named *Huanghao* or *Huanghuahao, Artemesia annua*), *yingzhao*

(*Artabotrys uncinatus*), *Herba agrimoniae*, and *Polyalthia nemoralis*, for example, were selected from the Chinese medicinal herbal heritage and developed with modern medical research techniques and methods through isolation of active monomers and identification of chemical structure. Some were even synthesized and underwent chemical structure modification. Besides *Qinghao* and *qinghaosu*, active monomers had been isolated from the herbs *Yingzhao* (*Artabotrys uncinatus*) and *Herba agrimoniae*, all with high antimalarial effects and with new chemical structures. But because their antimalarial effects were inferior to *qinghaosu*, or due to toxic side effects, scant material resources, or unfeasibility of extracting large quantities, they were not developed further.

Mosquito repellents were also discovered from this research into both chemical synthesis and Chinese medicinal herbs. Most of the products that were finally developed were from compounds found in the Chinese medicinal herbs, identified during the general population investigation and herb collection phase of Project 523.

Acupuncture for preventing and treating malaria was also a Chinese traditional method. It had been tested in many patients with a limited therapeutic effect that was thought to be the result of increasing the patients' immunity. But it was not effective in falciparum malaria patients who were non-immune or in those with a high parasitemia and persistent fever.

In synthetic drug research, more than ten thousand compounds were designed and synthesized, and more than forty thousand chemical substances were screened. For clinical studies, twenty-nine compounds (including drug combinations) were selected, and fourteen of these passed an evaluation. Lumefantrine and piperaquine were two examples. Both were slow to act but were long-acting. Artemisinin derivatives such as artemether and artesunate were fast-acting, highly effective, but with a high recrudescent rate. The combination of artemether and lumefantrine or of artemisinin and piperaquine not only maintained the benefits but overcame the weak points of both component drugs, and it also achieved the "one plus one equals more than two" augmented effect. These two combination drugs, in which one component was developed from Chinese traditional medicinal herbs and the other component was synthesized chemically, exemplified the process of combining Chinese traditional medicine and Western medicine to arrive at the same destination by different paths.

***

E. Unified and Selfless Contributions.

Project 523 came at the time when the Indochina War was escalating, China intensified its border defenses, and the ten-year turmoil and violence of the Cultural Revolution were underway. Project 523 was a scientific research mission, a military mission, and also a political mission. Under the guidance and concern of the Communist Party and national leaders, and the full attention and support of leaders at all subordinate levels, the scientific-technological research teams were highly motivated. Those involved considered Project 523 to be a way to express their patriotism and international support for an ally. The scientific and technological researchers felt proud and honored to be able to participate in this important mission. Project 523 became a great spiritual force to unify all participating and collaborative research teams. This positive spirit was transformed into a sense of responsibility for everyone, so they brought to the work not only intelligence, knowledge, and skills but also their passion. From laboratories in cities to clinical trials in the countryside and in the mountainous areas, everyone responded in spite of many difficulties, to achieve the stated goal.

Experts were sent into the war zone to study the clinical characteristics of falciparum malaria and its effect on the troops, and to observe how rapidly developed antimalarial drugs acted when used in the field. In 1966 the Academy of Medical Science assigned Ren Deli, Tian Xin, and Zhou Yiqing to follow the North Vietnamese troops travelling along the "Ho Chi Minh Trail" many times. They completed their clinical observations and wrote a report on falciparum malaria in Southeast Asia, making suggestions for treatment and prevention. In 1976, responding to the invitation of Cambodia, the National Ministry of Health again sent Zhou Yiqing together with Li Guoqiao, Shi Linrong, and others as an observation group to help Cambodia prevent and treat malaria. The group worked and lived in the high malaria endemic areas for six months; some members contracted falciparum malaria while others suffered from dengue fever. Yet they all insisted on continuing their studies on malaria and clinical trials of *qinghaosu* (artemisinin)-based new antimalarial drugs. Their attitude and work ethic were highly regarded and praised by the Cambodian authorities.

Many scientific and technological personnel worked for Project 523 in spite of many difficulties. During the Cultural Revolution, when violent conflict between two political groups broke out in Chongqing, the researchers of the Sichuan Institute of Chinese Materia Medica had to move to the basement to continue their experiments. At Shanghai Second Military Medical University, Project 523 researchers of the two opposing political groups criticized each other during the day but worked together at night. In Shanghai District, after supporters of the Cultural Revolution gained control, all research units were paralyzed. To avoid the interruption of Project 523, the local 523 office, after discussing with research teams, moved all researchers in synthetic antimalarial drug research to the Shanghai Institute of Parasitic Diseases, and all researchers for chemical mosquito repellants to Shanghai Pharmaceutical Factory No.2, to continue with the research project. The leader of the Military Academy of Medical Science gathered dozens of Project 523 researchers from its two institutes, and placed them under the direct leadership of the national Project 523 head office so they could not participate in their own unit's Cultural Revolution activities. In some units in the Nanjing District, Project 523 researchers were hassled by the opposition group. The Nanjing local 523 office, in spite of the political pressure, visited and persuaded each unit about the importance of Project 523, and requested that no conflicts or cessation of work be allowed without the agreement of the local 523 office.

To complete Project 523, many researchers willingly underwent serious hardships and experienced severe physical fatigue. In order to find the most effective anatomic sites for acupuncture or to test the effectiveness of a new drug, Li Guoqiao from Guangzhou College of Traditional Chinese Medicine, Yu Dechuan from the Beijing Institute of Materia Medica, and Guan Bizhen from Guangdong People's Hospital infected themselves with *Plasmodium falciparum* parasites. They tolerated the unpleasant symptoms and signs of malaria, including the high fever associated with the disease. Some research teams worked continuously for more than ten years in Hainan, on the Yunnan border, and in the poor mountainous malaria endemic areas, living with the villagers and leading a simple life with rough brown rice and plain vegetables. The 523 Chinese Herbs Investigation Team in Nanjing District walked sixty kilometers and climbed a thousand-meter-high mountain to investigate a Chinese regimen for treating malaria. Guangzhou Sun Yatsen University Professor Jiang Jingbo's daughter was drowned participating in the village farming brigade. Professor Jiang let his wife take care of the burial while he went on an essential working trip. For Niu Xinyi of the Beijing Institute of Materia Medica, her husband had just passed away, and in spite of her sorrow she led a team to Hainan Island to carry out a clinical study. From the same institute, to expedite the progress of Chinese medicinal herb research, senior expert and Professor Fu Fengyong brought small animals for drug testing and the necessary equipment with him to a farming village in Hainan. There he interviewed villagers, collected potentially effective regimens, started on-site extraction of the respective herbs, and also started screening for antimalarial effects. Some units glorified the spirit of collectivism, considering others' benefit before their own. In the period when it was possible technically to extract only a small amount of *qinghaosu*, the Shandong Institute of Chinese Traditional Materia Medica and Yunnan Institute of Materia Medica still supplied the extract to the Beijing Institute of Traditional Chinese Materia Medica and Shanghai Institute of Organic Chemistry, for their chemical structure studies. Liang Li and others of the Beijing Institute of Biophysics established the absolute configuration of the *qinghaosu* structure, a key finding, yet remained in the background, not pursuing personal acclaim. During that time, because there were many others in the institute needing to use the same equipment, Liang's group worked at night to utilize the time most efficiently. One member of the group worked very hard, became unwell, fainted in the laboratory, and unfortunately died. A casualty of overwork and stress, or a death from natural causes?

During the turmoil years of the Cultural Revolution, the 523 leading groups and 523 local offices of both Guangdong Hainan and Yunnan took major responsibility in administering actual work sites and managing the logistics of delivering supplies. Directors Song Weizhou and Zhao Mingzhen of the Department of Health of Guangzhou Military District, Director Xi Xiaguang, Deputy Director Wang Li of the Department of Health of Kunming Military District, Brigade Leader Ge Guangming of the Epidemic Prevention Brigade of Hainan Military District, and Deputy Director Liu Tungling from the Department of Health of Hainan Administrative District personally communicated with research workers in their areas. They made the necessary yearly arrangements for twenty research teams who were collecting traditional antimalarial regimens from the general population, or were carrying out drug testing during that time. The Hainan Military District sent a military doctor to accompany each 523 team working on site in Hainan Island to coordinate and ensure the administration, safety, and logistical needs of each team. The Hainan and Kunming local 523 offices chose test sites and arranged for transportation for the research teams.

It should not be forgotten that great encouragement was given to Project 523 by the leaders of related ministries, departments, and organizations. During the difficult time of the Cultural Revolution, when Minister Qian Xinzhong of the Ministry of Health was allowed to return, he listened to the reports on the progress of the work of Project 523 and gave very positive comments and encouragement. He said that in the turmoil of the Cultural Revolution, Project 523 had not only produced scientific results, but also protected scientific research units and intellectuals. During the Cultural Revolution, Director Tian Ye of the Drug Administration of the National Commission of Science and Technology personally inspected and supervised work in the laboratories, listened to work progress reports, and encouraged scientific and technological personnel. She Deyi of the Ministry of Chemical Industries was very supportive of Project 523. Besides taking care of the 523-related issues within his ministry, he was also concerned about the entire Project 523, and he supported and solved many problems and needs within the jurisdiction of his ministry. Leaders and personnel of institutes and organizations at the provincial, city, district, or military district level were all very supportive of Project 523. The troops, collective farms, local departments of preventive medicine, and local departments of health in the malaria endemic areas of Hainan and Yunnan supported and assisted the work and living requirements of the Project 523 scientific research teams. Their contributions should not and will not be forgotten.

With the support and concern of Project 523 leaders at the various levels, the administrative personnel for all Project 523 local offices were loyal in their duty in the face of difficulties. They often gave up their own professions to join Project 523 and struggled for more than ten or even thirty years. In difficult and crucial times of Project 523 during *qinghaosu* research and development, with their sense of responsibility, dedication, and intelligence, they analyzed, coordinated, and gave guidance in solving all problems in a timely manner in order to advance the project. And yet when one looks back on this important discovery of *qinghaosu* and the many other discoveries that resulted, their names are not mentioned anywhere. The administrative personnel of the Project 523 local offices were either military personnel or civilians, and the majority of them remained on the task for more than ten years; they were hard-working and uncomplaining. Many of them, because of their long-term absence from their own units, were forgotten by the leaders of their institution, and were not considered at times of new work opportunities and promotions. This therefore affected their living standard and rank, and their associated benefits to a certain extent were lost. In Shanghai District, because of the large number and heavy workload of Project 523 research teams, the administrative personnel of the Shanghai 523 local office had a harder job to organize and coordinate. Local Office Director Wang Wansen had a car accident on his field trip and broke two ribs. All these administrative personnel worked hard, had no concern for their personal benefits, and made their own contribution, even in difficult times, towards helping our ally and keeping our troops combat ready. They were proud that they had not wasted these years and had no complaints and no regrets.

CHAPTER 10

Qinghaosu's (Artemisinin's) Discovery and Development: A Successful Coordinated Effort

The discovery of *qinghaosu* (artemisinin) was a collective effort by Chinese medical, scientific, and technical research personnel, and depended on the management capabilities of a central organization for coordinating the necessary resources. Many diverse activities and functions had to be assigned to appropriate task force units throughout the country. The challenges to be faced, and then solved, were given to those units that were assumed to have the ability to accomplish the assignment, as in a major military campaign. If an individual research unit was unable to solve the problem in the allotted time, the work was handed on, as is the baton in a relay race, to an alternative department. Former Deputy Minister of Health Huang Shuze wrote in his 1981 annual Project 523 summary report that, "Project 523's competence was shown by its ability to involve many institutions, covering large geographic areas, with close coordination of different tasks, and the linking of associated projects. It brought scientific and technological workers of many specialties together, combining field and laboratory work, and the production of needed materials, with the application and use of developed products in the clinical setting. Together they worked in an organized, pre-planned, and systematic way. This was the key for the success of Project 523's rapid and excellent results. Only with a large-scale collaboration of this nature was it possible to achieve such a success. The interchange of scientific experience and knowledge and the sharing of essential equipment were highly commendable; as was the policy of not keeping information secret, and all units assisting each other, when necessary, with information and suggestions."[52]

Professor Shen Jiaxiang, former chief engineer of the State General Administration of Medicine, member of the Chinese Academy of Engineering, and ex-member of WHO SWG-CHEMAL, participated in the *qinghaosu* international collaboration. He said that "Without Project 523 there would be no *qinghaosu*." Regarding Project 523 and the discovery of *qinghaosu* (artemisinin), this is a plain and blunt statement. If Vietnam had not requested China's help in managing drug-resistant falciparum malaria affecting military combat strength during the war against America, and if China did not also need to be prepared for war, then there would have been no need for Project 523, and no need to invest such a large amount of manpower and other resources to find and develop new drugs for this disease. Without these circumstances, no one would have considered undertaking the research to develop the Chinese medicinal herb *Qinghao*. It is hard to imagine that during those years of political turmoil, that individuals or units, without this specific task, would invest such large amounts of material and financial capital into research on *Qinghao* or *Huanghuahao*. Also, if it were not for the special characteristics of Project 523, there would not be such a strong, scientific, closely monitored, large-scale research collaboration model to follow. And the *qinghaosu* project would not have been able to move forward in times of difficulty and discouragement to the success it has today.

The discovery of artemisinin and its derivatives attracted a great deal of attention from medical professionals within and outside China. In March 1996, the Hong Kong QiuShi Scientific Foundation awarded the "Outstanding Science and Technological Achievement Collective Award," for the discovery of *qinghaosu* (artemisinin), to acknowledge the Chinese scientists' important contribution to mankind. The recipients included not only the units that discovered artemisinin but also the units that developed artemisinin derivatives. Recipient unit representatives included Tu Youyou, Li Guoqiao, Wei Zhenxing, Liang Juzhong, Zhou Weishan, Li Ying, Zhu Dayuan, Gu Haoming, Liu Xu, and Xu Xingxiang. In the award ceremony, former Minister of Health Chen Mingzhang introduced the Chinese collaborators to the chairman and guests. He congratulated the researchers for their intensive work ethic on the antimalarial drug front, and narrated stories about their achievements. He also reported that with the approval of the State Council, under the leadership of the Project 523 leading group and the 523 head office, the whole nation had labored under a unified plan. "It divided tasks and collaborated in the search for and development of antimalarial drugs that finally led to the birth of *qinghaosu*. The success of antimalarial *qinghaosu* research and development was the result of a long-term collaboration by many institutions and departments, and of multiple specialties in China. The ten scientific researchers receiving the awards today are only representing those who have made outstanding contributions to this project."[53] The Hong Kong QiuShi Scientific Foundation awarded the "Outstanding Achievement in Science and Technology Collective Award" to *qinghaosu* (artemisinin) and its derivatives as the research results of a collective group of scientists, and not to specific individuals.

In early 2004, King Bhumibol Adulyadej, representing Thailand, granted the Prince Mahidol Royal Award, Thailand's highest medical award, to Chinese medical, scientific, and technical workers for discovering *qinghaosu* (artemisinin) and its derivatives, and for the contribution of *qinghaosu* (artemisinin)-based combination drugs in preventing and treating malaria. The award bears the inscription "Collective Award for Chinese *Qinghaosu*."

Looking back on that unforgettable and proud period of history, while we are appropriately excited and proud of the individuals' and units' achievements in various research projects of 523, we should not forget those unknown heroes who have also contributed to the project. They have added glory to the pearl at the top of the pagoda but have remained as silent as the foundation stones at its base. The glory does not belong only to the named units or the few representatives who received the awards. "All results from Project 523 research are the result of the long-term collaboration of many institutions and departments and multiple units."[54] The results are the achievement not only of those named, but also of the silent contribution of those unnamed. The data seen in the 1978 Yangzhou *Qinghaosu* Research Evaluation Report is a summary of the labor of dozens of research units and hundreds of researchers. The unit names of eight of the fourteen presenters in this evaluation meeting were not even mentioned in the "*Qinghaosu* Certificate for Invention." For example, the Sichuan Institute of Chinese Materia Medica and the Jiangxu Gaoyou County Department of Health were considered as principal research units, but their names were replaced by others on the "National Certificate for Invention." Therefore those that accepted the honor should not forget this, and the historic contribution of the unnamed in *Qinghao* and *qinghaosu* research shall not be erased. Besides the units receiving awards, there were forty-nine collaborative units and hundreds of scientific and technological researchers whose devotion and contribution are acknowledged and respected.

The results of *qinghaosu* research fully express the devotion Chinese scientists give to their motherland; it demonstrates the high spirit and collectivism of many intellectuals and revolutionaries and their selflessness; and their dedication to the work, ignoring the loss of personal benefits. One should always remember that this spirit is the force leading the Chinese people to international prominence and prosperity.

***

The discovery of *qinghaosu* (artemisinin) is a socialist victory song written for China by the devoted and selfless Chinese scientists and cadres of the revolution. The discovery of *qinghaosu* (artemisinin) will enter the history books on medical development in China, as well as the books on medical development in the rest of the world.

REFERENCES

1. WHO, "Anti-malarial Drug Combination Therapy. Report of a WHO Technical Consultation," WHO/CDS/RBM/2001/35, 2001, reiterated in 2003.
2. Goodman and Gilman, *The Pharmacological Basis of Therapeutics*, 828–829.
3. Richard Haynes, *US Fareast Economy Review*, March 14, 2002.
4. Zhou Yiqing, "The Effects of Malaria on Military Actions in War History" (internal document), Beijing, March 1979
5. Institute of Medical Intelligence, Military Academy of Medical Science, *Military Medical Data outside China* 5, no. 6(1973).
6. National Commission of Science and Technology, and Logistics Department of the People's Liberation Army, minutes of "Antimalarial Drugs Research Collaboration Group Conference" and "Collaborative Protocol on Antimalarial Drug Research," Beijing, June 1967.
7. National Malaria Prevention and Treatment Research Leading Group Office (in future referred to as National Project 523 Office), "Malaria Research 1967–1980 Results" (internal document).
8. Ge Hong, *Zhou Hou Bei Ji Fang* (*A Handbook of Prescriptions for Emergencies*)ca. 340 AD.
9. Jiangxu Gaoyou County Department of Health, "A New Way of Treating Malaria with *Qinghao*," April 1977In previous National Project 523 Office, *Consolidation of Accumulated Data on Project 523 and Qinghaosu*, March 2004.

10. Project 523 Group, "Literature Extract: Chinese Traditional Medicine Recipes for Malaria," June 1970. See "Controversy on Ownership Rights Regarding Contributions to *Qinghaosu*'s Discovery 1994," in previous National Project 523 Office, *Consolidation of Accumulated Data on Project 523 and Qinghaosu*, March 2004

11. Gu Guoming, "A Memoir on Partial Participation in *Qinghaosu* Research," June 5, 2004.

12. National Project 523 Office, "Report on the Antimalaria Drug Research Specialty Group Meeting; and Request Solution to Problems Encountered," Office Document No.5, May 31, 1972.

13. Project 523 Clinical Study Group, Institute of Chinese Materia Medica, China Academy of Chinese Traditional Medicine, "Summary of Clinical Trial No. 91," October 1972. See "Controversy on Ownership Rights Regarding Contributions to *Qinghaosu*'s Discovery 1994," in previous National Project 523 Office, *Consolidation of Accumulated Data on Project 523 and Qinghaosu*, March 2004

14. Institute of Chinese Materia Medica, China Academy of Traditional Chinese Medicine, *Research on Antimalarial Effects of Qinghao 1971–1978*, 26.

15. Ibid., 27.

16. National Project 523 Office, "Progress on *Qinghao* Antimalaria Research," October 1977.

17. Juye Prevention Team No.3, Shandong Institute of Parasitic Diseases, "Treating Vivax Malaria Patients with *Huang* No.1: A Preliminary Clinical Observation," September 27, 1973. See *Shandong Institute of Chinese Materia Medica's Research Data (Special issue on Huanghuahao antimalarial research [internal data])* 12 January (1980): 55.

18. Yunnan Institute of Materia Medica, "Preliminary Pharmacology and Toxicology Studies of *Huanghaosu*," October 1973. See Yunnan District Project 523 Office *Accumulated Data on Malaria Prevention and Treatment*

Research: Special Issue on Huanghaosu, February (1977):9–17.

19. Yunnan Institute of Materia Medica, "Preliminary Chemical Study of *Huanghaosu*," October 1973. See ibid. 4–8.

20. National Project 523 Office, "Meeting of Leaders of All 523 Offices: Summary Report," January 17, 1974. In previous National Project 523 Office, *Consolidation of Accumulated Data on Project 523 and Qinghaosu*, April 2004.

21. National Project 523 Office, "Summary Report on the *Qinghao* Research Meeting," in *Summary Reports on Malaria Prevention and Treatment Research* 1, April (1974).

22. Previous National Project 523 Head Office, "Controversy on Ownership Rights Regarding Contributions to *Qinghaosu*'s Discovery 1994," in *Consolidation of Accumulated Data on Project 523 and Qinghaosu*, March 2004.

23. Shandong Institute of Chinese Traditional Medicine, "Treating Vivax Malaria Patients with Simple Preparations of *Huanghuahaosu* and *Huanghuahao* Acetone Extract: A Preliminary Observation," in *Data on Chinese Traditional Medicine Research (Special issue on Huanghuahao antimalaria research [internal data])* 12 January (1980):57.

24. Yunnan District *Huanghaosu* Clinical Verification Group, 523 Team of Guangzhou College of Chinese Traditional Medicine, "Treating 18 Malaria Patients with *Huanghaosu*: A Summary," February 1975, in previous National Project 523 Office, *Consolidation of Accumulated Data on Project 523 and Qinghaosu*, April 2004.

25. National Project 523 Office, "523 Research Plan for 1975," in ibid.

26. Daily work records of Mr. Zhang Jianfang, director of National Project 523 Office, February 20, 1975.

27. "Ministry of Health Representative's Conversation and Statements after Listening to the Summary Report of the

Meeting for all 523 Local Offices Leaders" (transcribed from notes), March 5, 1975.

28. Beijing Institute of Traditional Chinese Materia Medica. Research Briefings Volume 3; June 26, 1975. (Reports sent to leader of China Academy of Traditional Chinese Medicine; copied to National Project 523 Office.)

29. 523 Traditional Chinese Medicine Group, "523 Traditional Chinese Medicine Research Plan for 1975," attachment to "523 Traditional Chinese Medicine Specialty Group Meeting: Summary Report," April 24, 1975.

30. Ministry of Health, Ministry of Petroleum and Chemical Industry, General Logistics Department of Chinese People's Liberation Army, Chinese Academy of Sciences, "Major Missions and Requests for the Last 3 Years of the 5-Year Antimalaria Research Plan," attachment to "Progress of the Antimalaria Research Project: Meeting Summary," July 1973.

31. Wu Yulin, letter to the National Commission of Science and Technology, September 24, 1994. See "Controversy on Ownership Rights Regarding Contributions to *Qinghaosu*'s Discovery in 1994," in previous National Project 523 Office, *Consolidation of Accumulated Data on Project 523 and Qinghaosu*, March 2004.

32. Zhan Eryi, "Main Aspects of Yunnan Institute of Materia Medica's *Qinghaosu* Research," Kunming, April 29, 2004.

33. Record Office, Shanghai Institute of Organic Chemistry of the Chinese Academy of Sciences, *Laboratory Raw Data Workbook*, September 3, 1975. See "Controversy on Ownership Rights Regarding Contributions to *Qinghaosu*'s Discovery in 1994," in previous National Project 523 Office, *Consolidation of Accumulated Data on Project 523 and Qinghaosu*, March 2004.

34. National Project 523 Office, "Progress in Research in Treating Malaria with Chinese Medicinal Herb *Qinghao*," in National Project 523 Office, *Summary Reports on*

Malaria Prevention and Treatment Research 1, February 8 (1976).

35. Malaria Prevention and Treatment Observation Group to Democratic Cambodia, "Report of the Malaria Prevention and Treatment Observation Group to Democratic Cambodia," July 1976.

36. National Project 523 Office, "Research on Antimalaria Effect of *Qinghao*: A Summary Report," in National Project 523 Office, *Summary Reports on Malaria Prevention and Treatment Research* 3, September 7 (1976).

37. National Project 523 Office, Office Document No.1, February 1, 1977.

38. National Project 523 Office, "Report on the Technical Exchange and Training Course in Methods for Measuring *Qinghao* Contents." See National Project 523 Office *Summary Reports on Malaria Prevention and Treatment Research* 1, April 13 (1977).

39. National Project 523 Office "Combined Chinese Traditional Medicine and Western Medicine Specialty Group Meeting in Nanning on Antimalaria Drug Research." See National Project 523 Office, *Summary Reports on Malaria Prevention and Treatment Research* 2, June 20 (1977).

40. Secretariat for Malaria Prevention and Treatment Research Conferences, "Summary Reports on Malaria Prevention and Treatment Research Conference," March 26, 1977.

41. National Malaria Prevention and Treatment Research Leading Group, "Certificate of Evaluation and Validation of *Qinghaosu*," Yangzhou City, Jiangxu Province, November 28, 1978.

42. Sichuan District Project 523 Office, "Report on Scientific Verification of *Qinghaosu* Tablets." See Sichuan District Project 523 Office, *Research on Antimalaria Effects of Qinghao* (internal data) May 6, 1978,44.

43. State General Administration of Medicine, the Logistics Department of the People's Liberation Army, and the

National Ministry of Health, "Notice Designating Institutes for the Urgent Production of *Qinghaosu (Huanghaosu)*," January 23, 1979. In previous National Project 523 Office, *Consolidation of Accumulated Data on Project 523 and Qinghaosu (1967–1981)* March 2004

44. Guangxi Guilin Pharmaceutical Factory, Chinese Traditional Medicine Laboratory 523 Team, "Research on *Qinghaosu* Derivatives" (internal data), November 1978.

45. Guangxi Guilin Pharmaceutical Factory, "Recent Progress in *Qinghaosu*-Based Antimalaria Drugs," *Guangxi Medical Journal* 10, October 25 (2003).

46. "A New Generation of Antimalarial Drugs—The Research on Dihydro-Artemisinin and Dihydro-Artemisinin Tablets (A New Drug)," http://www.cintcm.ac.cn/catcm/zy/RCG.htm.

47. Song Shuyuan, Teng Xihe, Ding Linmu, Li Peizhong et al., "Research on the Acute Toxicity of *Qinghaosu* Oil Emulsion Preparation on Monkeys," *Chinese Journal of Pharmacology* 1 (1983): 21–23.

48. Huang Yunchang, National Ministry of Health, speech at a press conference on *qinghaosu* and its derivatives. See previous National Project 523 Office, *Consolidation of Accumulated Data on Project 523 and Qinghaosu (1981–1988)*, March 2004.

49. Li Guangjin et al., "Permission for Production of *Qinghaosu* and Its Derivatives," *ChineseJournal of Medical Pharmacology* September 24(1987). See ibid.

50. National Commission of Science and Technology and various ministries,"Notice on Expediting the Promotion, Exportation and Foreign Exchange Earnings of *Qinghaosu*-Based Antimalarial Drugs," attachment to "Minutes of the Meeting Discussing the Promotion,Exportation and Foreign Exchange Earnings of *Qinghaosu*-BasedAntimalarial Drugs," Document (88) No. 427, July 16, 1998.

51. *Qinghaosu* Directional Committee, "Progress on *Qinghaosu* and Its Derivatives Research Plan for 1983 and 1985." See previous National Project 523 Office, *Consolidation of Accumulated Data on Project 523 and Qinghaosu (1981–1988)*, March 2004.

52. Deputy Minister of Health Huang Shuze, speech at the meeting of directors of all local 523 leading group offices, March 5, 1981. See Previous National Project 523 Office *Consolidation of Accumulated Data on Project 523 and Qinghaosu (1967-1981.)* 2004.3.

53. Minister Chen Mingzhang, Ministry of Health, speech at the QiuShi Scientific Foundation Award Ceremony at Beijing Science and Technology Hall, August 30, 1996. See previous National Project 523 Office, *Consolidation of Accumulated Data on Project 523 and Qinghaosu (1988–1996)*, March 2004.

54. Secretariat of the Directional Committee for research on *qinghaosu* and its derivatives, *Summary Reports on Meetings Regarding Collaborative Research on Qinghaosu and Its Derivative* 1, January (1982). See previous National Project 523 Office, *Consolidation of Accumulated Data on Project 523 and Qinghaosu (1981–1988)*, March 2004.

CPSIA information can be obtained at www.ICGtesting.com
ted in the USA
02s1335170316

BV00001B/10/P